VIROLOGY RESEARCH PROGRESS

THE ZIKA VIRUS: BACKGROUND, ISSUES, AND U.S. RESPONSE CONSIDERATIONS

VIROLOGY RESEARCH PROGRESS

Additional books in this series can be found on Nova's website under the Series tab.

Additional e-books in this series can be found on Nova's website under the e-book tab.

VIROLOGY RESEARCH PROGRESS

THE ZIKA VIRUS: BACKGROUND, ISSUES, AND U.S. RESPONSE CONSIDERATIONS

VERONICA SALAZAR
EDITOR

nova publishers
New York

NOTICE TO THE READER

The Publisher has taken reasonable care in the preparation of this book, but makes no expressed or implied warranty of any kind and assumes no responsibility for any errors or omissions. No liability is assumed for incidental or consequential damages in connection with or arising out of information contained in this book. The Publisher shall not be liable for any special, consequential, or exemplary damages resulting, in whole or in part, from the readers' use of, or reliance upon, this material. Any parts of this book based on government reports are so indicated and copyright is claimed for those parts to the extent applicable to compilations of such works.

Independent verification should be sought for any data, advice or recommendations contained in this book. In addition, no responsibility is assumed by the publisher for any injury and/or damage to persons or property arising from any methods, products, instructions, ideas or otherwise contained in this publication.

This publication is designed to provide accurate and authoritative information with regard to the subject matter covered herein. It is sold with the clear understanding that the Publisher is not engaged in rendering legal or any other professional services. If legal or any other expert assistance is required, the services of a competent person should be sought. FROM A DECLARATION OF PARTICIPANTS JOINTLY ADOPTED BY A COMMITTEE OF THE AMERICAN BAR ASSOCIATION AND A COMMITTEE OF PUBLISHERS.

Additional color graphics may be available in the e-book version of this book.

Library of Congress Cataloging-in-Publication Data

ISBN: 978-1-53610-124-9

Published by Nova Science Publishers, Inc. † New York

CONTENTS

PREFACE

In late 2015, health officials in Brazil recognized a marked increase in the number of infants born with microcephaly (from Greek, meaning “small head”), a birth defect that may accompany significant, permanent brain damage. Although not conclusive, the increase in microcephaly is suspected to be related to the emergence of Zika virus infections in Brazil early in 2015. Zika virus is related to the viruses that cause yellow fever, dengue, West Nile, and Japanese encephalitis viruses. Historically, Zika virus was found in Africa. Since 2007, Zika transmission has also occurred in Southeast Asia and the Western Pacific. In the Americas, transmission was first identified in Brazil in May 2015. As of January 29, 2016, the Pan American Health Organization (PAHO) had received reports of local transmission in 25 countries and territories in the Americas, including Puerto Rico and the U.S. Virgin Islands. This book discusses scientific and technical aspects of Zika virus infection, including modes of transmission, symptoms, diagnosis and treatment, and prevention. Policy concerns and official actions are evolving and will also be discussed.

In: The Zika Virus
Editor: Veronica Salazar
ISBN: 978-1-53610-124-9

Chapter 1

ZIKA VIRUS: BASICS ABOUT THE DISEASE*

Sarah A. Lister

INTRODUCTION

In late 2015, health officials in Brazil recognized a marked increase in the number of infants born with microcephaly (from Greek, meaning "small head"), a birth defect that may accompany significant, permanent brain damage.[1] Although not conclusive, the increase in microcephaly is suspected to be related to the emergence of Zika virus infections in Brazil early in 2015.[2]

Zika virus is related to the viruses that cause yellow fever, dengue, West Nile, and Japanese encephalitis viruses.[3] Historically, Zika virus was found in Africa.[4] Since 2007, Zika transmission has also occurred in Southeast Asia and the Western Pacific. In the Americas, transmission was first identified in Brazil in May 2015. As of January 29, 2016, the Pan American Health Organization (PAHO) had received reports of local transmission in 25 countries and territories in the Americas, including Puerto Rico and the U.S. Virgin Islands.[5]

Symptoms of Zika virus infection are usually mild, and the illness often resolves quickly. However, the possible link to microcephaly has raised the level of public health concern in the Americas and elsewhere, and has spurred

---

* This is an edited, reformatted and augmented version of a Congressional Research Service publication R44368, prepared for Members and Committees of Congress, dated February 4, 2016.

wide-ranging efforts to protect pregnant women from infection. *Although some aspects of Zika virus infection (such as the behavior of the mosquitoes that carry it) are well understood, many other aspects of this emerging infection in the Americas are unknown.* While study of the Zika virus ramps up, health officials will have to base their decisions on the best available evidence, which could have substantial gaps for the time being.

On February 1, 2016, the World Health Organization (WHO) determined that the rapid spread of Zika infections in the Americas and the suspected link to a serious birth defect constituted a "Public Health Emergency of International Concern" under the International Health Regulations.[6] Under this designation, nations are expected to cooperate, expand monitoring, and share information in order to stem the spread of the disease.

This report discusses scientific and technical aspects of Zika virus infection, including modes of transmission, symptoms, diagnosis and treatment, and prevention. Policy concerns and official actions are evolving and will be discussed in forthcoming CRS products. Unless otherwise cited, information in this report is drawn from the Zika virus information pages of the U.S. Centers for Disease Control and Prevention (CDC),[7] WHO,[8] and PAHO,[9] where readers can also find the most current public health recommendations for the general public, travelers, and pregnant women.

Transmission

The primary means of Zika virus transmission is through the bite of an infected mosquito. Zika infections are not commonly transmitted directly between people, although the following direct transmission routes have been reported: from an infected pregnant woman to the fetus, through sexual contact with an infected person, and through blood transfusion from an infected blood donor. Zika virus has been detected in breast milk of infected mothers, but transmission via breast milk has not been reported.[10]

Zika virus is carried by and transmitted through the bite of a specific genus of mosquito, *Aedes*, primarily *A. aegypti* (see *Figure 1*), and possibly *A. albopictus* (also known as the Asian tiger mosquito). Both species have noticeable black and white stripes. These are the same species that transmit the viruses that cause yellow fever, dengue, and chikungunya. As a result, the behavior, geographic ranges, and control methods for these mosquitoes are generally well understood. Neither species is native to the Americas. Both were unintentionally introduced. Their geographic ranges vary from year to year, but in the United States they are limited to warmer regions. *A. aegypti* is found in the South and parts of the Southwest. The range of *A. albopictus* is broader, including parts of the central United States, Ohio Valley, and Eastern Seaboard as far north as New York City. Both species are aggressive daytime biters, well adapted to urban environments.[11]

Source: James Gathany, CDC Public Health Image Library, image #9261, http://phil.cdc.gov.

Figure 1. The *Aedes aegypti* mosquito, the most common vector of Zika virus.

Health officials are most concerned about areas where Zika virus *transmission* has been reported, rather than where human infections have been reported. Because *Aedes* mosquitoes are present in the United States, CDC says that" ... [w]ith the recent outbreaks [in Central and South America], the number of Zika virus disease cases among travelers visiting or returning to the United States will likely increase. These imported cases may result in local spread of the virus in some areas of the United States. CDC has been monitoring these epidemics and is prepared to address cases imported into the United States and cases transmitted locally."[12]

The WHO says that an *animal reservoir* for the Zika virus is "unknown."[13] An animal reservoir is a non-human animal that also harbors a pathogen and contributes to its spread. (Examples are birds and West Nile virus, deer and the Lyme disease bacteria, and bats and Ebola virus.) Research on possible animal reservoirs of Zika virus is underway. If an animal reservoir is present, it could amplify the number of infected mosquitoes and increase the risk of transmission to humans.

Symptoms

According to CDC, "About 1 in 5 people infected with Zika will get sick. For people who get sick, the illness is usually mild. For this reason, many people might not realize they have been infected. The most common symptoms of Zika virus disease are fever, rash, joint pain, or conjunctivitis (red eyes). Symptoms typically begin 2 to 7 days after being bitten by an infected mosquito."[14] Zika virus usually remains in the blood of an infected person for about a week, but it can persist for longer periods in some people.

WHO is concerned that the spike in cases of microcephaly in Brazil is consistent with the recent introduction of the Zika virus into populations that lack immunity to it. Health officials are investigating possible links between Zika infection and other neurological and congenital disorders, including Guillain-Barré syndrome (GBS), an immune disorder often associated with bacterial or viral infections, characterized by muscle weakness and sometimes paralysis.[15]

Diagnosis and Treatment

CDC reports that no commercially available diagnostic tests are available for Zika virus disease. Zika virus testing is performed at CDC and a few state health departments. Clinicians are advised to contact their state health department to facilitate testing.[16]

No specific therapy or vaccine against Zika infection is approved by the U.S. Food and Drug Administration (FDA) for use in the United States, nor is one available elsewhere in the world. For the time being, treatment of infected persons is supportive, including treating symptoms and maintaining hydration. CDC, the National Institutes of Health (NIH), and pharmaceutical companies are working to develop tests, potential treatments, and vaccines.[17] Officials caution that it is highly unlikely that a vaccine would be available in less than one or two years.

Preventing Transmission

Absent a vaccine, avoiding bites from infected mosquitoes is the primary means to prevent transmission.[18] According to WHO, "At present, the most important protective measures are the control of mosquito populations and the prevention of mosquito bites in at-risk individuals, especially pregnant women."[19] Official recommendations for areas in which infected mosquitoes may be present are (1) for everyone, routine use of protective clothing and repellants, and staying indoors with windows closed or screened; and (2) for pregnant women, consideration of postponing travel to affected areas, and diligent protection against mosquito bites if travel is unavoidable.[20]

End Notes

[1] Pronounced my-kroh-SEF-uh-lee. See http://www.merriam-webster.com/dictionary/micro cephaly. For more information see U.S. Centers for Disease Control and Prevention (CDC), "Facts about Microcephaly," http://www.cdc.gov/ncbddd/birthdefects/microcap haly.html.

[2] See CRS Insight IN10433, Zika Virus: Global Health Considerations, by Tiaji Salaam-Blyther.

[3] This paragraph is drawn from CDC, "Zika Virus Spreads to New Areas—Region of the Americas, May 2015–January 2016," MMWR, vol. 65, no. 3, January 29, 2016, http://www.cdc.gov/mmwr/volumes/65/wr/pdfs/mm6503.pdf.

[4] Zika virus was first isolated in 1947 from a monkey in the Zika forest of Uganda.

[5] PAHO is the Regional Office for the Americas of the World Health Organization (WHO). See http://www.paho.org.

[6] WHO, "WHO Announces a Public Health Emergency of International Concern," press release, February 1, 2016, http://www.who.int/en/. For more information on the International Health Regulations, see "International Health Regulations (IHR)" in CRS Report R43829, Ebola: Selected Legal Issues, by Jared P. Cole et al.

[7] CDC, "Zika Virus," http://www.cdc.gov/zika/index.html.

[8] World Health Organization (WHO), "Microcephaly/Zika virus," http://www.who.int/emer gencies/zika-virus/en/.

[9] Pan American Health Organization (PAHO), "Zika virus infection," http://www.paho. org/hq/index.php?option=com_content&view=article&id=11585&Itemid=41688.

[10] CDC, "Zika virus, Transmission," http://www.cdc.gov/zika/transmission/index.html.

[11] CDC, "Surveillance and Control of Aedes aegypti and Aedes albopictus in the United States," http://www.cdc.gov/chikungunya/resources/vector-control.html.

[12] CDC, "Zika Virus Disease Q & A: Should We Be Concerned about Zika in the United States?" http://www.cdc.gov/zika/disease-qa.html.

[13] WHO, "Zika virus fact sheet," January 2016, http://www.who.int/mediacentre/factsheets/ zika/en/.

[14] CDC, "Zika Virus Disease Q & A," January 29, 2016, http://www.cdc.gov/zika/disease-qa.html.

[15] WHO, "WHO statement on the first meeting of the International Health Regulations (2005) (IHR 2005) Emergency Committee on Zika virus and observed increase in neurological disorders and neonatal malformations," February 1, 2016, http://www.who.int/mediacentre/news/statements/2016/1st-emergency-committee-zika/en/.

[16] CDC, "Zika Virus for Healthcare Providers," http://www.cdc.gov/zika/hc-providers/index.html.

[17] Comments of Dr. Tony Fauci, Director of the NIH National Institutes of Allergy and Infectious Diseases, Transcript for CDC Telebriefing: Zika Virus Travel Alert, January 28, 2016, http://www.cdc.gov/media/archives.htm.

[18] Mosquito control approaches are beyond the scope of this report. For more information on U.S. domestic mosquito control activities, see CDC, "Mosquito Control," http://www.cdc.gov/westnile/faq/mosquitoControl.html.

[19] WHO, "WHO Announces a Public Health Emergency of International Concern," press release, February 1, 2016, http://www.who.int/en/.

[20] CDC, "Zika Virus Disease Q&A: What can people do to prevent becoming infected with Zika?" http://www.cdc.gov/zika/disease-qa.html; and CDC, "Interim Guidelines for Pregnant Women during a Zika Virus Outbreak—United States, 2016," MMWR, vol. 65, no. 2, January 22, 2016, http://www.cdc.gov/mmwr/volumes/65/wr/mm6502e1.htm.

In: The Zika Virus
Editor: Veronica Salazar
ISBN: 978-1-53610-124-9

Chapter 2

ZIKA VIRUS IN LATIN AMERICA AND THE CARIBBEAN: U.S. POLICY CONSIDERATIONS*

Clare Ribando Seelke, Tiaji Salaam-Blyther and June S. Beittel

SUMMARY

Congress is debating how to respond to an ongoing outbreak of Zika virus, a mosquito-borne illness that has no treatment or vaccine and can cause microcephaly—a severe birth defect—and other neurological complications. As of June 16, 2016, 60 countries and territories had reported mosquito-borne transmission of the virus, 39 of which are in Latin America and the Caribbean and are reporting cases of Zika for the first time. Brazil, which has registered the most confirmed cases of Zika in Latin America, will host the summer Olympics in August 2016. Scientists expect that travel destinations in the Caribbean will see more cases as the summer's warm, rainy season continues. More than 750 U.S. citizens, including pregnant women, have become infected through either travel or sexual transmission.

Frequent business and tourist travel, combined with the close proximity and similar climates of Latin America and the southern United States, means that mosquito-borne Zika infections are likely in the United

* This is an edited, reformatted and augmented version of a Congressional Research Service, Publication No. R44545, dated June 29, 2016.

States. Zika is primarily spread by *Aedes* mosquitoes—primarily *Aedes aegypti* but also *Aedes albopictus*, the latter of which is present in a majority of U.S. states. Local (or mosquito-borne) transmission has not yet occurred in the continental United States but is occurring in Puerto Rico and the U.S. Virgin Islands.

On February 8, 2016, the Obama Administration submitted an emergency request for almost $1.9 billion in supplemental funding to respond to the Zika outbreak, including $526 million for international efforts. On April 6, 2016, the Administration announced that it would reprogram $589 million in unobligated funds, including $510 million in Ebola supplemental funds, for efforts to address the Zika outbreak. The U.S. Agency for International Development (USAID) is reprogramming $215 million of that funding—including a $78 million transfer to the U.S. Centers for Disease Control and Prevention (CDC)—for international efforts. In mid-May 2016, both the House and the Senate passed supplemental appropriations measures for Zika response. The House bill, H.R. 5243, would provide $622.1 million in Zika funding and rescind an equal amount of budget authority. The Senate measure (S.Amdt. 3900 to H.R. 2577, the combined FY2017 Military Construction-Veterans Affairs and Transportation-Housing and Urban Development appropriations bills) would provide $1.1 billion in Zika response funding without rescissions. On June 23, 2016, the House agreed to a conference agreement (see H.Rept. 114-640) that would provide $1.1 billion for Zika response, including $175.1 million for State Department and USAID activities. On June 28, 2016, the Senate voted not to invoke cloture on the conference agreement.

The number of people in the Western Hemisphere affected by Zika is unknown, but as many as 4 million people may be at risk of infection in 2016, and nearly all countries in Latin America and the Caribbean have recorded cases of the virus. Zika responses in the region have been led by Brazil and Colombia, multilateral organizations such as the World Health Organization (WHO)/Pan American Health Organization (PAHO), and the U.S. government. Health experts have expressed concerns about the capacity of health systems—particularly in Central America and the Caribbean—to prevent, diagnose, and care for Zika cases and associated complications, particularly among pregnant women. Related issues of interest to Congress include how to balance support for U.S. initiatives and multilateral approaches, the proper scope and components of U.S. health assistance to the region, and funding for pandemic preparedness and research on neglected tropical illnesses in Latin America.

INTRODUCTION

As of June 16, 2016, the World Health Organization (WHO) reported that 60 countries worldwide had experienced mosquito-borne transmission of Zika, 46 of which had never had Zika cases before (see **Figure 1**).[1] The U.S. Centers for Disease Control and Prevention (CDC) has concluded that Zika causes microcephaly (a serious birth defect involving brain damage) and is associated with Guillain-Barré syndrome (GBS, a neurological condition) and other neurological and autoimmune conditions. Latin America and the Caribbean have been most affected by this outbreak (**Figure 1**). Since February 2016, new cases in the southern parts of the region have decreased, but an increase in new cases is expected in the Caribbean during the summer months as mosquitoes hatch and bite and as the United States will likely experience an increase in travel-associated cases and possibly local transmission (**Figure 2**).[2] Policymakers are concerned about the spread of Zika into the continental United States, as well as the potential that visitors traveling to the Olympic Games in Brazil in August could contract the virus and bring it back to their home countries.

In February 2016, WHO Director-General Margaret Chan announced that the International Health Regulations (2005) Emergency Committee on Zika virus had determined that the Zika outbreak was a Public Health Emergency of International Concern (PHEIC).[3] Shortly thereafter, President Barack Obama requested that Congress provide almost $1.9 billion in emergency appropriations to fund domestic and international responses to the outbreak. A conference agreement on Zika funding was approved by the House but remains pending in the Senate. As Congress considers funding the Zika request and then exercises oversight over U.S. Zika responses in Latin America and the Caribbean, Members may consider issues such as the following:

- **Balance between U.S. bilateral and multilateral Zika responses.** Although U.S. health assistance (bilateral and regional) to Latin America in general has declined, U.S. support for the Pan American Health Organization (PAHO)[4] has increased. While considering the President's Zika request and the FY2017 budget request, Congress may discuss how much support to provide for multilateral responses led by PAHO and for bilateral efforts.
- **U.S. health programs in Latin America as a part of U.S. policy toward the region.** On average, roughly 10%-20% of the funds

provided by the State Department and the U.S. Agency for International Development (USAID) for Latin America between FY2009 and FY2016 have been aimed at health programs. The WHO has recommended that women and men in countries with local transmission[5] of Zika be correctly informed and oriented to consider delaying pregnancy.[6] Congress may consider what levels of health funding would be appropriate given the WHO recommendation and the limited access to sexual education and contraception in the region.

- **Regional apportionment and components of global health budget.** Less than 5% of all U.S. global health funds are provided to Latin America, and the vast majority of these funds are for HIV/AIDS programs. On average, health indicators in the region—particularly those related to maternal and child health, family planning, and reproductive health—are better than in other low- and middle-income countries, although inequities exist.[7] Congress might reexamine the apportionment of global health funding and consider whether investments in the region are sufficient to meet emerging health concerns.
- **Funding for pandemic preparedness.** The United States committed to support 30 countries and the Caribbean Community (CARICOM)[8] in bolstering their pandemic preparedness through the Global Health Security Agenda (GHSA).[9] Two of the 30 countries are in Latin America and the Caribbean (Haiti and Peru). Congress included almost $600 million in emergency Ebola appropriations to the CDC in support of GHSA and $50 million in FY2016 appropriations to USAID for pandemic preparedness activities. Congress might consider funding levels for those programs and where those funds are allotted.
- **Investments in research and development of neglected diseases.** In recent years, *Aedes* mosquitoes have caused three disease outbreaks (dengue, chikungunya, and Zika) in Latin America and the Caribbean, all of which have been imported into the United States, with the latter being only travel-associated at the time of this report. These and other diseases lack vaccines to prevent transmission, treatment regimens, and effective vector control tools. Congress might evaluate options to address threats from new and reemerging diseases, including those that are mosquito-borne.

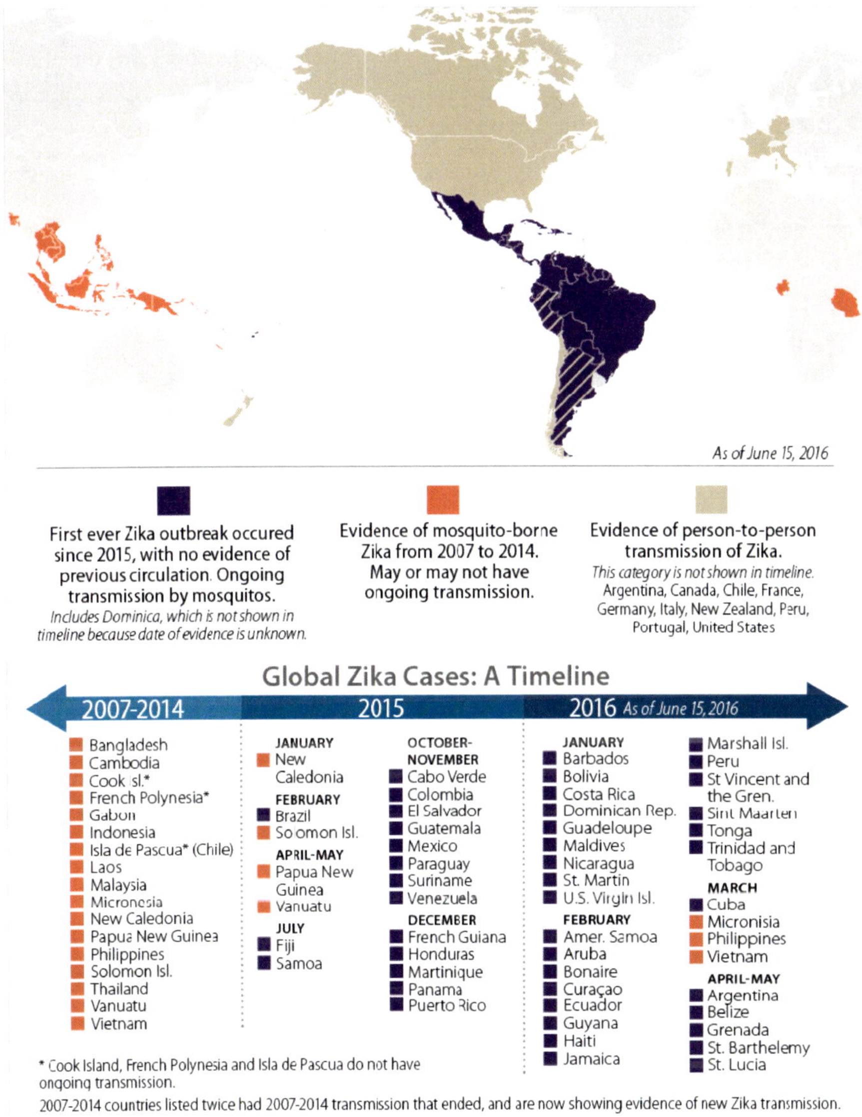

Source: Adapted by the Congressional Research Service (CRS) from World Health Organization (WHO), *Zika Virus, Microcephaly, Guillain-Barré Syndrome*, June 16, 2016.

Notes: Zika virus is not necessarily present throughout the countries/territories shaded in this map.

Figure 1. Global Transmission of Zika: 2007-2016.

This report provides background information on the Zika virus, discusses challenges faced by governments and implementing partners in the Latin America and Caribbean region that are attempting to control the ongoing outbreak, and analyzes the above issues in the context of the U.S. Zika response.

BACKGROUND

Zika was discovered in the Zika forest of Uganda in 1947. From that time until 2007, when the first large Zika outbreak was recorded, Zika virus infection primarily caused mild symptoms (fever, skin rash, conjunctivitis, muscle and joint pain) that resolved within one week.[10] In 2007, the first large Zika outbreak was recorded on the Micronesian island of Yap. Household surveys detected 185 cases.[11]

Scientists are studying the virulence of the Zika virus and the extent to which human activity affects global spread.[12] Retrospective studies of a 2013-2014 outbreak in French Polynesia linked Zika infection with GBS for the first time.[13] The current outbreak, which began in Brazil, has been accompanied with a spike in microcephaly and GBS cases, as well as other neurological and autoimmune disorders.[14] As of June 16, 2016, WHO has reported more than 1,600 cases of Zikarelated microcephaly worldwide—almost all in Brazil. In addition, 13 Zika-affected countries have recorded an increased incidence of GBS.[15]

Aedes-Related Outbreaks in Latin America and the Caribbean

In the last century, Latin American and Caribbean countries have transformed from largely rural to mostly urban societies, with some 80% of all people in Latin America and the Caribbean now living in urban areas. Throughout the region, millions of people live in densely populated urban slums and poor rural communities where homes and other facilities lack air conditioning or window screens. Lack of proper plumbing and poor sanitation facilitate mosquito breeding, as mosquitoes can lay their eggs in standing water.[16] *Aedes* mosquitoes thrive in such conditions, biting during the day and breeding indoors and out.[17]

Health threats to Latin American and Caribbean populations may also be exacerbated by the 2015-2016 El Niño weather pattern, which is reflected in unusually warm water in the eastern equatorial Pacific Ocean. The present El Niño phenomenon, which has been particularly strong, has produced multiyear droughts in some areas (Colombia, Venezuela, and northern Central America) and extreme flooding in others (Argentina, Uruguay, and Paraguay). The warm, wet weather has facilitated the proliferation of mosquitoes, and human responses to drought conditions have provided favorable conditions for mosquito breeding because more people have been storing water.[18] Studies have also linked climate change with greater health threats, such as increasing prevalence of malaria, chikungunya, and dengue fever.[19] In recent years, *Aedes* mosquitoes have spread three disease outbreaks across the Americas: chikungunya,[20], dengue,[21] and Zika. All three of these diseases have been imported into the United States and can become locally transmissible because the *Aedes* mosquito resides in large segments of the United States (see **Figure 2**).

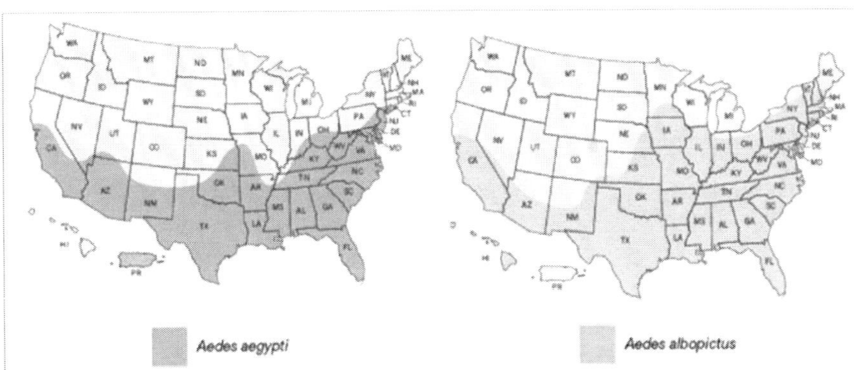

Source: CDC, "Estimated Range of *Aedes aegypti* and *Aedes albopictus* in the United States, 2016 Maps," http://www.cdc.gov/zika/vector/range.html, accessed on May 23, 2016.

Notes: Maps represent an estimate of the potential range of the *Aedes* mosquito. They are not intended to represent risk for spread of Zika.

Figure 2. Estimated Range of *Aedes* Mosquito in the United States.

Zika in Latin America and the Caribbean

Scientists are unsure how many people have been infected by Zika in the Western Hemisphere, but as many as 4 million people may be at risk of infection, and nearly all countries have recorded cases.[22] As of June 16, 2016, Brazil had 159,914 suspected Zika cases, almost 40,000 of which have been confirmed through diagnostic testing (see **Figure 3**).[23] Since the outbreak began in Colombia, the country had recorded 82,935 suspected cases, more than 8,000 of which have been confirmed.[24] Two key factors complicate efforts to count Zika cases:

1. About 75% of infected people do not develop symptoms.
2. The virus is detectable for less than seven days in infected people's blood.[25]

Zika: From Latin America to the United States and U.S. Territories

The Zika Virus Outbreak in the U.S.Territories[28]

Although Puerto Rico and the U.S. Virgin Islands are U.S. territories, when it comes to the Zika virus, their locations, climate, and ecology put them in company with much of Latin America and the Caribbean. PAHO is tracking the spread of the Zika virus in Puerto Rico, the U.S. Virgin Islands, and countries across Latin America and the Caribbean. On June 2, 2016, PAHO commented, "A downward trend of cases of Zika virus disease in Central and South America continues to occur while in most Caribbean countries and territories the trend continues to rise."

As of June 15, 2016, none of the 50 U.S. states had identified local transmission of Zika virus. However, local transmission was first identified in Puerto Rico in December 2015 and in the U.S. Virgin Islands in January 2016. Puerto Rico has now identified more than 1,300 laboratory-confirmed cases, including more than 140 pregnant women. Puerto Rico is working with the CDC to actively monitor these women and assure that they and their babies receive the best possible pre- and post-natal care. Full-term babies from the earliest of these pregnancies will be due in mid- to late summer. More than 20 locally acquired cases of the Zika virus have been identified in the U.S. Virgin Islands. Puerto Rico's Zika outbreak comes amid a fiscal emergency in the territory. The Obama Administration has requested assistance for Puerto Rico and the other territories as part of emergency supplemental appropriations for the national and global response to the Zika outbreak. Congress is considering an appropriations package at this time.

Sources and additional information: CRS Report R44460, Zika Response Funding: Request and Congressional Action; CRS Report R44275, Puerto Rico and Health Care Finance: Frequently Asked Questions; CRS Report R44095, Puerto Rico's Current Fiscal Challenges; CDC, "Zika Virus," http://www.cdc.gov/zika/; PAHO, "Zika Virus Infection," http://www.paho.org/zika.

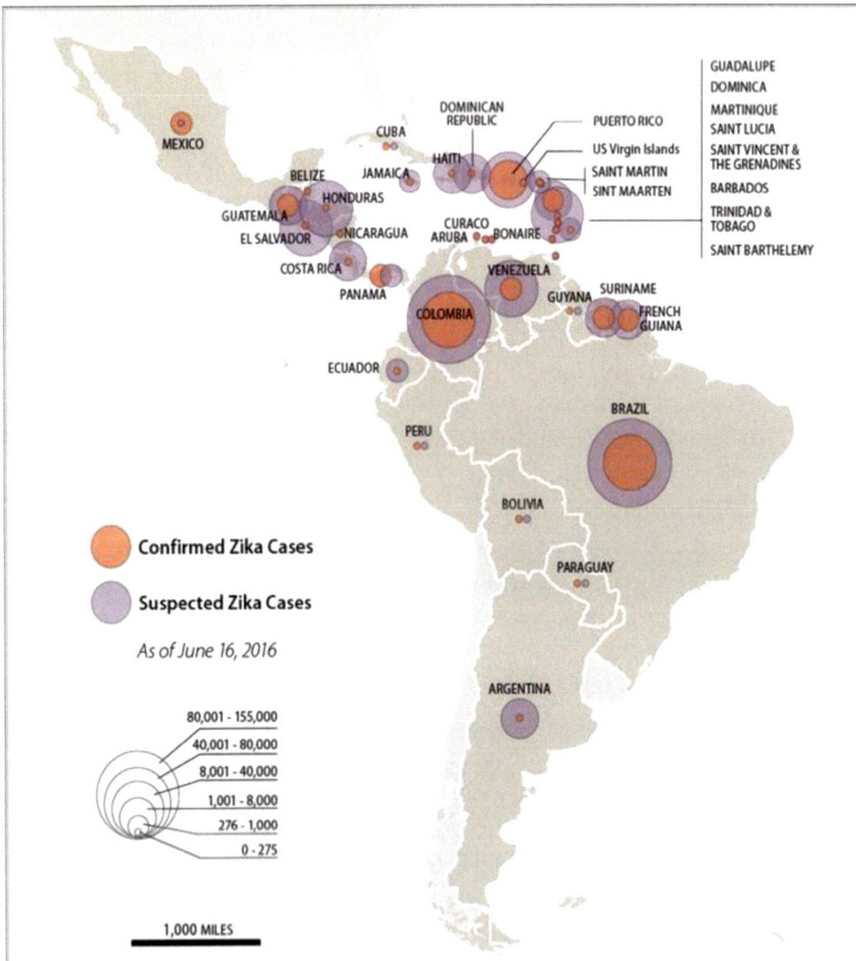

Source: Adapted by WHO/PAHO, Cumulative Zika Suspected and Confirmed Cases Reported by Countries and Territories in the Americas, June 17, 2016.

Figure 3. Confirmed and Suspected Zika Cases by Country; (as of June 16, 2016).

As of June 16, 2016, all Zika cases detected in the continental United States[26] (755) have been either acquired abroad or sexually transmitted, although the U.S. Virgin Islands, American Samoa, and Puerto Rico have experienced local transmission by mosquito. CDC and other health experts are preparing for the likelihood that the continental United States may experience locally acquired Zika cases this summer.[27] Given the broad range of the *Aedes* mosquitoes (see **Figure 2**) and the fact that mosquito-borne diseases have been imported into the United States previously, a successful response to the Zika outbreak may require U.S.-Latin American cooperation in surveillance, research, and response over several years.

International Zika Responses in Latin America and the Caribbean

Country Efforts

The number of Zika cases, the capacity of health systems to address them and related complications, and the plans to do so vary widely across Latin America and the Caribbean. Haiti, for example, lacks a functioning hospital system, and Venezuela has little capacity to provide basic maternal and child health care at this time. In terms of preparedness and response to epidemiological emergencies, including Zika, recent assessments of core capacities under the International Health Regulations carried out by PAHO have highlighted weaknesses in health system capacity in all Caribbean countries (including Suriname and Guyana), selected countries of Central America (El Salvador, Guatemala, Honduras, and Nicaragua), Bolivia, and Paraguay. The Inter-American Development Bank (IDB) shares this view.[29]

Experts are concerned that these countries and several others are much less equipped than Brazil and Colombia—two of the six countries in the Americas that PAHO deemed capable of handling a pandemic illness in 2012—to address Zika and related health consequences.[30] Inadequate laboratory and diagnostic capacity, poor access to sexual education and contraception, and resistance to national mosquito control efforts due to mistrust of government authorities have hindered efforts in some countries. Gang violence and insecurity have also reportedly prevented health workers from providing services in some parts of El Salvador and Honduras.

Should Zika-associated cases of microcephaly become more common outside of Brazil, the health systems in the region may come under strain. Most countries in Latin America and the Caribbean lack the capacity to treat children born with severe birth defects and do not generally permit abortion.[31] Lifetime care for a child with microcephaly can be expensive. In the United States, such care can cost up to $10 million.[32] Brazil has struggled to care for infants with microcephaly, many of whose families live hours from one of the few hospitals that can provide care.

Brazil[33]

Brazil has been at the epicenter of the ongoing Zika outbreak, with 159,914 suspected Zika cases since the beginning of the current outbreak, of which 39,993 have been confirmed as of June 16, 2016.[34] On November 11, 2015, Brazil's Ministry of Health declared a Public Health Emergency of National Importance in response to a sharp increase in the number of infants born with microcephaly. Whereas fewer than 200 cases of microcephaly were reported annually in Brazil prior to 2015, Brazil's Ministry of Health detected 1,581 microcephaly cases between the start of 2015 and June 16, 2016.[35]

In Brazil, as in most countries in Latin America, a diagnosis of microcephaly in utero does not meet the government's standards under which abortion is permissible.[36] As a result, Brazil's Ministry of Health has issued guidelines for providing physical and occupational therapy to children born with microcephaly and is certifying hospitals capable of providing care to those infants. Brazil's congress passed a law to provide a small monthly stipend to families caring for microcephalic children, many of whom are led by single mothers who have lost the ability to maintain employment due to the type of care microcephalic babies require. Many observers are concerned that there may be a rise in illegal abortions (and possibly maternal mortality due to unsafe abortions) in Brazil as a result of increasing diagnoses of microcephaly.[37]

The Brazilian government has launched a National Plan to Combat the *Aedes* Mosquito and Microcephaly, which includes research, prevention, and mosquito control efforts, as well as health assistance for pregnant women and children. Brazil has several world-class research institutions with vast experience in tropical diseases, and the country's national public health institutions are working with local and international partners to develop more efficient diagnostic kits, antiviral drugs, and a Zika vaccine.[38] The government has dispatched 220,000 troops and 300,000 health agents to communities around the country to educate the population and eliminate mosquito breeding grounds. Officials have placed particular focus on mosquito-control efforts in Rio de Janeiro, which is scheduled to host the 2016 Summer Olympic Games in August (see text box).

2016 Summer Olympics in Rio de Janeiro, Brazil, and the Zika Outbreak

Ongoing debate concerns whether the arrival of hundreds of thousands of tourists to Rio de Janeiro, Brazil, in August for the summer Olympics will hasten the global spread of the Zika virus. For some time, Brazilian officials have sought to assuage the fears of athletes and fans by pointing out that the Olympics will be occurring during the dry, winter season in Brazil, when fewer mosquitoes are present, and that all the venues and hotel areas will be regularly fumigated in preparation for the games. Regardless of those assurances, some health officials have called for the Olympics to be postponed or called off entirely. They have expressed concern that the strain of Zika present in Brazil, which has been linked to microcephaly and neurological problems, has been exported to Cape Verde, Africa. In May 2016, 150 health experts and bioethicists wrote an open letter to the WHO director-general urging her to recommend postponing the Olympics. In addition, some widely known athletes have indicated that they will skip the Rio Olympics due to concerns about the Zika virus.

Others oppose the proposal, noting that nonpregnant travelers going to Brazil can take sensible precautions to avoid mosquito bites and, upon returning home, use mosquito repellant (to avoid infecting mosquitoes) and use condoms to prevent sexual transmission of Zika. WHO made earlier statements indicating that "cancelling or changing the location of the 2016 Olympics will not significantly alter the international spread of Zika virus." On June 14, 2016, the WHO Emergency Committee on Zika met and reaffirmed its decision that "there is a very low risk of further international spread of Zika virus as a result of the Olympic and Paralympic Games as Brazil will be hosting the Games during the Brazilian winter when the intensity of [local] transmission of arboviruses, such as dengue and Zika viruses, will be minimal." U.S. epidemiologists calculated that the Olympic visitors would account for only 0.25% (i.e., less than 1%) of the total risk for spreading Zika through air travel.

Sources: Lena H. Sun, "150 Experts Say Olympics Must Be Moved or Postponed Because of Zika," *Washington Post*, May 27, 2016; WHO, "WHO Statement on the Third Meeting of the International Health Regulations (2005) (IHR(2005)) Emergency Committee on Zika Virus and Observed Increase in Neurological Disorders and Neonatal Malformations," June 14, 2016; Reuters, "New Research Finds Low Risk of Zika Virus at Olympics,' June 7, 2016.

Some observers have expressed concerns about the adequacy of Brazil's efforts, particularly in low-income areas. Although Brazil has one of the most advanced public health systems in Latin America, significant gaps in prenatal care and birth outcomes exist in the poor north and northeast regions, where many Afro-descendants live (and many Zika cases have been concentrated) compared to the wealthier south.[39] Reports also indicate that many states in Brazil's northeast region ran out of mosquito larvicide last year and that the country's fiscal challenges and political instability have inhibited efforts to address the Zika outbreak this year.[40]

Colombia

Colombia has the second-most cases of reported Zika virus infection in the Western Hemisphere. As of June 16, 2016, Colombia had reported 82,935 suspected cases, roughly 8,000 of which have been confirmed.[41] Two-thirds of Colombia's municipalities have reported suspected or confirmed cases, and Colombia's National Institute of Health estimates that between 200,000 and 300,000 people may contract the disease in the country by the end of 2016.[42]

Similar to Brazil, Colombia has a relatively sophisticated public health system. Over the past several years, Colombia has spent a little over $15 million annually on combating contagious pathogens. It plans to maintain that funding level in 2016.[43] The Colombian Ministry of Health issued a Zika virus risk-based preparedness and response plan in January 2016 that included four key elements:

1. Strengthening the national system of epidemiological surveillance;
2. Training health personnel on early detection, diagnosis, and management of Zika cases;
3. Coordinating Zika awareness, prevention, and response activities; and
4. Bolstering health care services to improve capacity to address Zika cases and related illnesses and to implement guidelines for comprehensive care of patients.

In January 2016, the Colombian Minister of Health visited the main cities around the country to raise awareness about Zika and build support for countering the disease among local health officials. The Ministry of Health also released policy recommendations that advised couples in affected areas to use contraceptive methods to prevent possible sexual transmission of the virus and postpone pregnancy. The government allotted an additional $1.4 million to purchase necessary supplies and improve institutional support to prevent and combat Zika.

As of June 16, 2016, Colombia had seven confirmed cases of microcephaly.[44] Some experts predict that Colombia may see an increase in microcephaly cases in the coming months because its Zika outbreak began roughly six months after Brazil's. Others maintain that numbers may not grow significantly because abortion is legal under certain circumstances in the country, and women may opt to abort after a prenatal microcephaly diagnosis.[45] Colombian officials predict that about 300 Zika-linked microcephaly cases may be diagnosed between May and September 2016.[46] The government is working with the U.S. CDC to monitor and treat women infected with the Zika virus.

Multilateral Organizations

WHO and PAHO

In 2015, Brazil experienced an unusual spike in microcephaly cases. Evidence later emerged linking Zika with microcephaly. The phenomenon prompted WHO Director-General Margaret Chan to convene an emergency committee on Zika virus to discuss four key issues:

1. The association of Zika infection with birth malformations and neurological syndromes;
2. The potential for further international spread of the virus given the wide geographical distribution of the mosquito vector;
3. The lack of population immunity in newly affected areas; and
4. The absence of vaccines, specific treatments, and rapid diagnostic tests.[47]

Upon recommendations of the emergency committee, Chan declared that the Zika outbreak was a PHEIC in February 2016. A PHEIC declaration signals that the health event may require immediate international action and often prompts a coordinated, multinational response. Also in February, the WHO released a Strategic Response Framework and Joint Operations Plan to guide the international response to the outbreak, neonatal malformations, and neurological conditions associated with the virus from January to June 2016. The plan focused on the following:

- Strengthening disease surveillance,
- Building laboratory capacity to detect the virus,
- Bolstering mosquito control,
- Providing care for infected persons, and
- Defining and supporting priority research areas.[48]

The plan asked for donors to provide $25 million to WHO (to coordinate and support global responses to the outbreak and scientific studies on the virus) and PAHO (to coordinate and support implementation of responses in the Americas). The funds were intended to support national activities related to disease surveillance; responses to the Zika outbreak, microcephaly, and GBS; and research from January through June 2016. The remaining $31 million in funds was requested for partner organizations, including the U.N. Children's Fund (UNICEF) and U.N. Population Fund.[49]

Due to sluggish contributions, in May 2016, U.N. Secretary General Ban Ki-moon announced the establishment of a U.N. Zika Response Multi-Partner Trust Fund to attract support for unfunded priorities outlined in the aforementioned WHO strategic framework. Contributions to the fund can be made by U.N. member states, regional organizations, nongovernmental organizations, businesses, and individuals. As of June 22, 2016, donors had provided more than $4 million (16%) to the WHO/PAHO portion of the request.[50] In addition, WHO released $3.8 million in emergency support for the Zika response. WHO and PAHO do not count the emergency funds toward fundraising goals, as these funds are to be reimbursed.

On June 16, 2016, WHO issued a revised $122 million Zika Strategic Response Plan to guide international efforts from July 2016 through December 2017.[51] The plan describes how the WHO/PAHO and 60 partner organizations aim to bolster detection, prevention, care and support, and research on Zika and related complications. It prioritizes support for women of childbearing age and their partners in communities affected by Zika. In addition, the plan urges countries and donors to bolster investments in counseling, reproductive health services, abortions (where legal), and postnatal follow-up and care for women who have been infected with Zika and for children born with microcephaly. It also calls for expanding services and research on GBS.

Since 2015, PAHO has been working with the U.S. CDC, USAID, National Institutes of Health (NIH), U.S. Department of Defense (DOD), and other leading research entities in the region to fill in significant knowledge gaps about Zika, its transmission, and its complications and to develop new diagnostic tests and hasten progress toward creating a vaccine. PAHO has partnered with governments, U.N. entities, multilateral development banks, and private organizations on disease surveillance, mosquito abatement, community engagement and education campaigns on personal protection, and services for those affected. PAHO, the World Bank, and the IDB have developed an assessment tool that measures country capacity to handle anticipated Zika cases and accompanying complications.[52]

World Bank

In recent years, the World Bank has scaled back many of its health programs in Latin America to focus on regions with greater health needs, particularly sub-Saharan Africa. Nevertheless, the World Bank has maintained health strengthening projects in Argentina, Brazil, Nicaragua, and El Salvador and is launching a new project in Panama.[53] It is also analyzing countries to forecast the impact of major events, such as Zika, on economies in affected countries. In February 2016, the World Bank initially estimated that the costs of the Zika outbreak to Latin America would be moderate at $3.5 billion, or roughly 0.06% of regional gross domestic product.[54] Officials have since stated, however, that the impact could be greater, particularly for Caribbean countries dependent on tourism from non-Zika affected countries.[55]

Since the Zika outbreak began in Latin America, the World Bank has made $150 million available for assistance. Most of the funding would be made available by restructuring existing projects to include components focused on Zika or by including Zika-related activities in new projects. Nine countries in Latin America qualify for International Development Association[56] aid based on their low per-capita incomes. These countries can therefore access a "crisis response" window that would provide new money (including grants or credits). As of mid-June 2016, Guyana had asked for $5 million from the crisis response window.[57] El Salvador had allocated $4 million from an existing health project, and Nicaragua had allocated $1 million from an existing project for Zika responses. Two states in Brazil had requested $20 million. Another large effort with the federal government of Brazil had been put on hold due to the political challenges unfolding in the country.

Inter-American Development Bank

The IDB has an active portfolio of loans and grants for health programs in Latin America totaling roughly $2.7 billion. The largest programs are in Brazil, Mexico, the Dominican Republic, El Salvador, Nicaragua, and Panama. Most IDB programs focus on health system strengthening to enable governments to improve the provision of maternal and child health care and to better address noncommunicable diseases. The IDB is considering a request from the Caribbean Health Agency and a joint proposal from four South American countries to help improve those countries' compliance with International Health Regulations (2005)[58] and assist in planning for health emergencies.

In response to the Zika outbreak, the IDB has offered to reorient up to $180 million of its current portfolio of water, sanitation, and health programs to address Zika.[59] Of the $60 million in health financing made available, the IDB had received requests to reorient $19 million as of mid-May 2016. IDB efforts to address Zika are generally focused on providing family planning through community health workers to rural areas, training primary care workers to detect nervous-system problems, and distributing supplies to prevent mosquito bites. The IDB is also supporting communications campaigns and efforts to increase surveillance capacity and vector control. In addition, the IDB has partnered with New York University to launch a crowdsourcing project that would enable governments to seek and partner with global health experts for responses to Zika and other infectious disease outbreaks.

U.S. Government

Supplemental Request

In February 2016, President Barack Obama requested almost $1.9 billion in emergency supplemental funding to address the Zika outbreak (**Table A-1**), the bulk of which was requested for the Department of Health and Human Services (HHS) primarily for domestic response. The Administration requested that $376 million of those funds be used by USAID and the Department of State for international responses and $150 million be used for CDC international responses.

Reprogrammed Funds for USAID and CDC Programs

On April 6, 2016, the Administration announced that it would reprogram $589 million in unspent Ebola funds to address the Zika outbreak. USAID is reprogramming $215 million of that funding to support short-term efforts in Latin America, including a transfer of $78 million to CDC for Zika activities in the region and a $4 million transfer to the Department of State for a contribution to the International Atomic Energy Agency. The largest components of USAID funds are to support vector control programs ($50 million) and to fund a grand challenge to encourage innovative responses to vector control, diagnostics, surveillance, and personal protection ($30 million). By far the largest component of USAID's transfer to CDC is to support surveillance, epidemiology, and public health studies ($44 million) (See **Table A-2**).

USAID and CDC are determining where to program their activities based on the anticipated numbers of cases in each country (based on experience with dengue and chikungunya) and the anticipated needs of countries in the region. Central America and the Caribbean are top priorities. Unlike the supplemental request, which included $10 million for operating costs, the reprogrammed funds will be implemented with existing staff and resources (although USAID may use program funding to hire some staff).[60] The funds are expected to last for eight months to a year at most.[61]

Congressional Action on the Budget Request[62]

In mid-May 2016, both the House and the Senate passed supplemental appropriations measures for Zika response. The House bill, H.R. 5243, would provide $622.1 million in Zika funding and rescind an equal amount of budget authority. The Senate measure (S.Amdt. 3900 to H.R. 2577, the combined FY2017 Military Construction-Veterans Affairs and Transportation-Housing and Urban Development appropriations bills) would provide $1.1 billion in Zika response funding without rescissions. On June 23, 2016, the House agreed to a conference agreement (see H.Rept. 114-640) that would provide $1.1 billion for Zika response, including $175.1 million for State Department and USAID activities. On June 28, 2016, the Senate voted not to invoke cloture on the conference agreement.

ISSUES TO CONSIDER

Congress is considering a range of domestic and international responses to the Zika outbreak. In the global context, as summarized above (see "Congressional Action on the Budget Request"), Members are debating the appropriate response to the outbreak.[63] In the international context, Congress may consider how to balance support for U.S. bilateral and multilateral Zika responses. U.S. foreign assistance to Latin America and the Caribbean has been declining since FY2011, and USAID has been phasing out many global health programs in the region. The Zika outbreak may prompt broader discussions about whether to bolster U.S. global health investments in the

Western Hemisphere, including in reproductive health. Discussions about controlling the Zika outbreak may also focus on U.S. support for global pandemic preparedness efforts, as well as research and development for diagnostics, treatments, and prevention measures for certain neglected diseases.

Balancing Support for Multilateral and Bilateral Zika Responses

U.S. bilateral and regional health assistance for Latin America and the Caribbean has declined by roughly 34% during the Obama Administration. In contrast, U.S. annual assessed contributions to PAHO have increased from $59.1 million in FY2009 to $65.7 million in FY2015. U.S. voluntary and assessed contributions currently represent roughly 37% of the organization's $200 million annual budget.[64] The U.S. government also provides annually assessed and voluntary contributions to WHO, UNICEF, the World Bank, and the IDB, which have launched Zika responses in the region. PAHO has struggled to fund missions that bring experts to the region, which cost between $15,000 and $20,000 per expert. PAHO maintains that those missions, partnerships with NIH and other research entities, and training for health and vector control workers are greatly needed in the region.[65]

WHO and PAHO have launched a revised plan to support countries in their response to the Zika outbreak (see "WHO and PAHO," above) through 2017. As of mid-June 2016, WHO and PAHO have received $4 million to support what they estimate will cost $122 million overall.[66] The WHO has not reported voluntary contributions from the U.S. government for the plan, although the Administration's February 2016 Zika budget request includes $10 million for the WHO/PAHO response and the Administration reprogrammed $14 million of Ebola funds for voluntary contributions to WHO, PAHO, and UNICEF for Zika activities in April 2016.[67]

Given the relatively small U.S. health investment in the region (as discussed below), Congress may consider providing funds for the WHO/PAHO Zika plan. Congressional discussions may also identify the appropriate mix of funds, if any, to provide for bilateral and multilateral Zika responses. In a region with relatively deep engagement with multilateral organizations, some would argue for providing increased funds to those entities rather than investing bilaterally. Others believe that it is harder to control how multilateral contributions are spent and ensure that related activities align with U.S. priorities.

Health as a Component of USAID Development Assistance to Latin America and the Caribbean[68]

Current U.S. policy in Latin America is designed to promote economic and social opportunity, ensure the safety of the region's citizens, strengthen effective democratic institutions, and secure a clean energy future. As part of broader efforts to advance these priorities in the region, USAID and the State Department provide foreign assistance. USAID funding in Latin America and the Caribbean has been declining, particularly since FY2011—the period following the release of the 2010 President's Policy Directive on Global Development (PPD-6).[69] Although the directive sought to elevate development as a "core pillar" of American foreign policy, it also directed U.S. agencies to "be more selective about where and in which sectors [they] work."

The PPD-6 prompted USAID to conduct a comprehensive review of its assistance programs. Following a review of the health sector, USAID determined that many of the countries in the region "had achieved remarkable progress, were far ahead of other presence countries [i.e., those with a USAID mission], and could effectively sustain progress without further USAID assistance."[70] This finding led to a gradual reduction in health funding in the region (**Figure 4**). Health assistance fell from a high of $445.6 million in FY2010 to an estimated $231.3 million in FY2016.

USAID continues to support programs related to maternal and child health, reproductive health, and family planning in Guatemala and Haiti, but it has discontinued such programs in all other countries across the region.[71] Policymakers concerned about development in Latin America may consider whether health should once again become a larger component of U.S. assistance programs, as inadequate access to health services can exacerbate poverty and inequality. A recent report by the U.N. Development Program urged governments and donors in the region to focus on addressing exclusion experienced by Afro-descendant and indigenous populations, women who suffer from domestic violence, and rural populations.[72] These issues may be of particular concern in Central America, where the Administration significantly increased development assistance in FY2016 (and in the FY2017 request) but did not include any funds for bilateral health programs in Honduras or El Salvador.

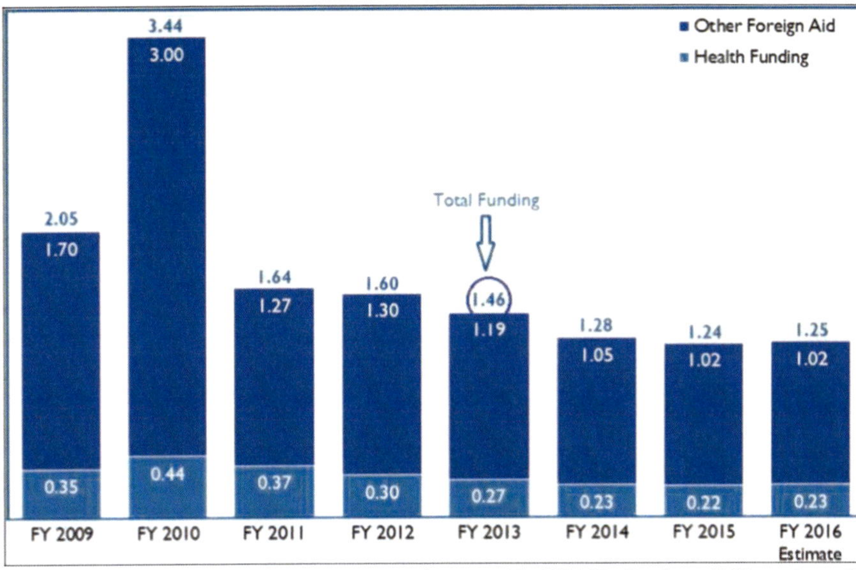

Source: Created by CRS from correspondence with USAID Budget Office, June 8, 2016.

Figure 4. Department of State and USAID Assistance to Latin America and Caribbean: FY2009-FY2016; (current U.S. $ in millions).

U.S. Global Health Support in Latin America

With support from USAID and other donors, many Latin American countries have made notable progress in improving the delivery of primary health care services—including access to contraception and basic prenatal care—over the past few decades. Since the 1960s, these developments have led to a 41% decline in maternal mortality, a 70% decline in infant mortality, and a drop in fertility rates from six children per family to between two and three children per family.[73]

As a result of this progress, USAID global health engagement in the region has been steadily declining. In FY2009, for example, USAID implemented global health programs in 14 countries in the region. By FY2016, USAID global health programs were operating in five countries (Brazil, Dominican Republic, Guatemala, Guyana, and Haiti). In three of those countries (Brazil, Dominican Republic, and Guyana), USAID global health programs focused only on HIV/AIDS by FY2016.[74] In FY2009, maternal and child health (23%) and reproductive health and family planning programs

(13%) together constituted almost 40% of USAID health programs in Latin America. By 2016, maternal and child health (8%) and family planning and reproductive health programs (7%) amounted to 15% of all USAID health spending in the region.

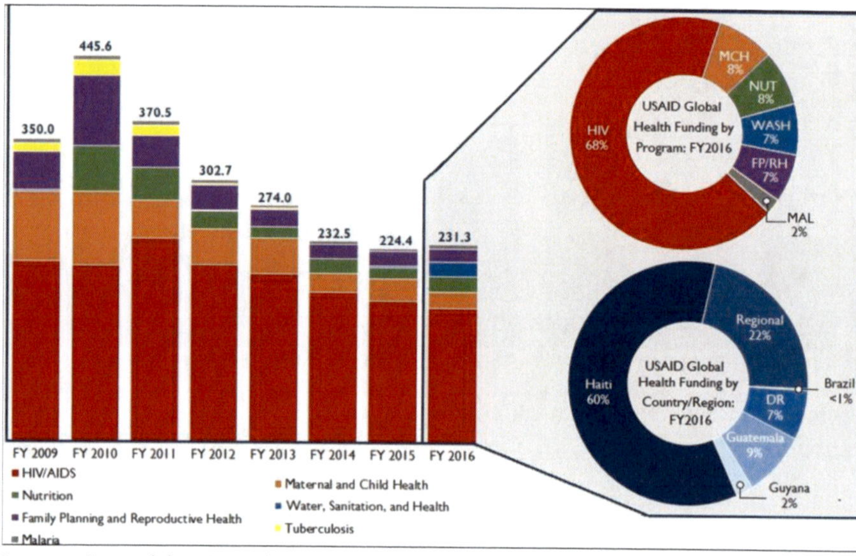

Source: Created by CRS from correspondence with USAID Budget Office, June 8, 2016.

Notes: FY2009-FY2015 are enacted levels and FY2016 is an estimate. MCH = maternal and child health; NUT = nutrition; WASH = water, sanitation, and hygiene; FP/RH = family planning/reproductive health; MAL = malaria; DR = Dominican Republic; Regional = regional programs.

Figure 5. USAID Global Health Funding in Latin America and the Caribbean: FY2009-FY2016; (current U.S. $ in thousands and percentage).

In 2016, health ministers in several countries urged women to consider postponing pregnancy in response to Zika. Delaying or avoiding pregnancy is a problem in some communities where access to affordable and reliable contraception is limited and domestic violence (including spousal rape) is a major problem.[75] Experts estimate that some 56% of pregnancies in Latin America are unplanned, particularly among adolescents in poor communities who often do not have access to sexual education or counseling.[76] Health experts, including WHO Director-General Margaret Chan, have decried this constraint.[77] The WHO/PAHO Zika Strategic Response Plan, Revised for July 2016-December 2017 urges donors to prioritize providing counseling, reproductive services, and pre- and postnatal care to pregnant women and their partners who may be affected by Zika and related birth defects.

Some groups have advocated for many countries in Latin America and the Caribbean to legalize abortions for severe microcephaly cases and have urged the United States to bolster investments in family planning and reproductive health services across the region.[78] Others oppose the legalization of abortion and argue that additional resources should be spent on finding effective treatments and vaccines.

Pandemic Preparedness

In recent years, a succession of new and reemerging infectious diseases have caused outbreaks and pandemics that have together affected millions of people worldwide: Severe Acute Respiratory Syndrome (2003), Avian Influenza H5N1 (2005), Pandemic Influenza H1N1 (2009), Middle East Respiratory Syndrome coronavirus (2013), Ebola (2014), and Zika (2015). In 2014, former HHS Secretary Kathleen Sebelius and WHO Director-General Margaret Chan announced the Global Health Security Agenda (GHSA), a global effort to accelerate implementation of the International Health Regulations (2005), particularly in resource-poor countries that lack the capacity to comply with the regulations.[79] The regulations describe measures to strengthen global capacity to respond to public health events with potential international impact.

Some analysts have asserted that Zika transmission has exposed weaknesses in Latin American countries' pandemic preparedness.[80] As of 2012, PAHO maintained that only six countries (Brazil, Canada, Colombia, Costa Rica, and the United States) were prepared to handle a pandemic.[81] Through GHSA, the United States has committed to support 30 countries—two of which (Haiti and Peru) are in the Western Hemisphere—and CARICOM.[82]

Congress provided $597 million to CDC through emergency Ebola appropriations for the GHSA. The legislation did not specify where the funds were to be used. In light of the growing disease threats posed by *Aedes* mosquitoes in the Western Hemisphere, policymakers might consider whether to expand U.S. support for GHSA implementation to other countries in Latin America.

Given entreaties from health experts to bolster pandemic preparedness efforts to minimize the effects of future outbreaks, Congress might consider whether ongoing funding for pandemic preparedness is sufficient. Since FY2014, Congress has been appropriating $72.5 million annually to USAID and an average of $58 million annually to CDC for global health security efforts. The Administration is seeking $72.5 million for USAID and $65.2 million for CDC to implement global health security programs in FY2017.

Research and Development for Neglected Diseases

Several U.S. government agencies are supporting development of vaccines and treatments for the Zika virus.[83] DOD is conducting preclinical research on a Zika vaccine candidate and plans to start human testing by the end of 2016. Under its innovations Broad Agency Announcement, the Biomedical Advanced Research Development Authority (BARDA) within HHS has received a number of proposals to develop new Zika vaccines using vaccine platforms that could also be used for other emerging infectious disease threats. BARDA, NIH, and DOD are collaborating to support the development of a vaccine at DOD's Walter Reed Army Institute of Research. Initial small-scale development is proceeding, and preclinical evaluation will begin soon.

The WHO and other health experts have called for increasing investments in research and development for "neglected tropical diseases" such as dengue and chikungunya, as well as for Zika.[84] A consortium of health experts estimates that the international community would need to double current investments in health research and development for neglected diseases from $3 billion in 2014 to $6 billion by 2020 to meet global health goals.[85]

According to the Global Health Technologies Coalition, the United States accounts for roughly 70% of public investment and 45% of global investment in global health research and development.[86] The United States provides the highest amounts of funding for research and development for 26 of the 30 most neglected diseases. As of 2012, it supported more than half of the global health products in the development pipeline. Since peaking in 2009, however, U.S. funding for global health research and development has fluctuated (Figure 6).

In April 2016, Congress enacted P.L. 114-146, Adding Zika Virus to the FDA Priority Review Voucher Program Act, to add the Zika virus to the list of tropical diseases eligible for the Food and Drug Administration's (FDA) Priority Review Voucher program, which allows companies to fast-track a product through the FDA regulatory process. Congress has not yet enacted legislation, however, that would provide additional funding for Zika-related research. Some analysts recommend that Congress revise authorizing language to ensure that products granted an FDA priority voucher are affordable.[87]

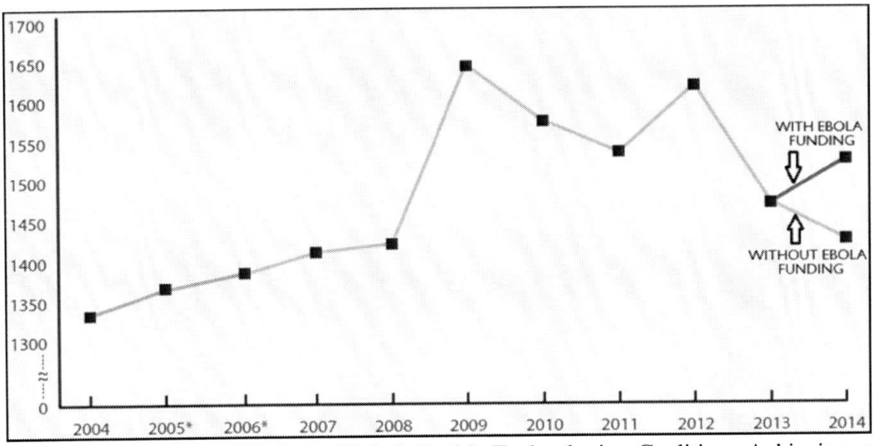

Source: Adapted by CRS from Global Health Technologies Coalition, Achieving a Bold Vision for Global Health: Policy Solutions to Advance Global Health R&D, 2016, p. 6.

Notes: Estimates. Data sources: Global Health Technologies Coalition, "Saving Lives and Creating Impact: Why Investing in Global Health Research Works," Policy Cures; 2012; and M. Moran et al., "Neglected Disease Research and Development: The Ebola Effect," G-FINDER, Policy Cures, 2015.

Figure 6. U.S. Investment in Research and Development for Neglected Diseases; (2014 US$ in millions).

Health experts also argue that the Zika outbreak highlights the need to increase research into effective vector control measures. At the 69th World Health Assembly, WHO Director-General Chan asserted that "the spread of Zika, the resurgence of dengue, and the emerging threat from chikungunya are the price being paid for a massive policy failure that dropped the ball on mosquito control in the 1970s."[88] Inappropriate and inconsistent use of insecticides has led to insecticide resistance and the growth of *Aedes* populations across the Western Hemisphere.[89] This phenomenon has prompted some to advocate for increased U.S. investments in vector control in the region. Brazil has become the first country in the world to approve the large-scale use of genetically modified mosquitoes in vector control programs, and the FDA is reviewing a request to conduct trials of the technology in the United States.[90]

The Zika budget request includes $100 million for USAID to incentivize the development of Zika vaccines, diagnostics, and vector control measures. It is unclear whether these funds will be sufficient to encourage market development of these products.

POSSIBLE FUTURE ACTIONS

The future of the Zika supplemental funding request is uncertain at this time due to a number of controversial measures that are in the conference report. Some observers are concerned that there may not be enough time for the House and Senate to resume negotiations to reconcile their responses to the Zika outbreak before they adjourn in July.[91] In the meantime, U.S. agencies are using reprogrammed funds to complement current efforts by Latin American governments, PAHO, the private sector, and other donors. The size and scope of U.S.-funded initiatives to address Zika in Latin America and the Caribbean is also uncertain.

APPENDIX A. SUPPORTING DOCUMENTATION

Table A-1. Zika Funding Request

Department/Agency	Key Activities	Funding Level
Department of Health and Human Services (HHS)		$1.509billion
U.S. Centers for DiseaseControl and Prevention (CDC)	• Support domestic Zika responses	$743 million, of which $150million is for international efforts
	• Enhance domestic mosquito control programs	
	• Establish domestic rapid response teams	
	• Improve domestic laboratory and surveillance capacity	
	• Expand the domestic CDC Pregnancy Risk Assessment Monitoring System, improve domestic Guillain-Barré syndrometracking, and ensure the ability of domestic birth defect registries to detect Zika-related risks	

Department/Agency	Key Activities	Funding Level
	• Increase research linking Zika virus infections and microcephaly	
	• Enhance international capacity for Zika surveillance, expand the Field Epidemiology Training Program (FETP), laboratory testing, health care provider training, and vector surveillance and control in countries at highest risk of Zika virus outbreaks	
	• Improve diagnostics for Zika virus and support developments for vector control	
Centers for Medicare and Medicaid Services	• Temporary one-year increase in Puerto Rico's Medicaid federal medical assistance percentage to support health services for pregnant women at risk of infection or diagnosed with Zika and for children with microcephaly and other related health costs	$250 million
Vaccine Research and Diagnostic Development and Procurement	• Research, rapid advanced development and commercialization of new vaccines and diagnostic tests for Zika virus	$200 million
	• Establish an Urgent and Emerging Threat Fund to address Zika and other outbreaks	$210 million
	• Support Puerto Rico's community health centers in preventing, screening, and treating Zika, expand home visiting services for low-income pregnant women at risk of Zika infection, and provide targeted maternal and child health services	
U.S. Agency for International Development (USAID)	• Implement vector management activities in countries at risk of Zika infection	$335 million
	• Stimulate private sector research and development of vaccines, diagnostics, and vector control innovations	
	• Support training of health care workers in affected countries	
	• Support for pregnant women's health	
	• Establish Zika education campaigns	
	• Issue a Global Health Security Grand Challenge calling for innovative diagnostics, vector control, personal protection, community engagement, and surveillance for Zika and other infectious diseases	
	• Provide flexibility in the use of remaining USAID Ebola funds	

Table A-1. (Continued)

Department/Agency	Key Activities	Funding Level
U.S. Department of State	• Support for U.S. citizens and State Department employees in affected countries, public diplomacy, communications, and other operations activities • Support World Health Organization and Pan American Health Organization efforts to minimize Zika threats in affected countries	$41 million

Source: White House, "Preparing for and Responding to the Zika Virus at Home and Abroad," February 8, 2016.

Notes: The FETP is CDC's program to train epidemiologists worldwide. For more information, see http://www.cdc.gov/globalhealth/healthprotection/fetp/index.htm.

Table A-2. Funding Redirected from USAID Ebola Emergency Operations for International Zika Responses (current US$ in millions)

Department/ Agency	Key Activities	Funding Level
USAID/U.S. Department of State	• Social and behavioral change communications	$17.0
	• Vector management	$50.0
	• Grand challenge for development	$30.0
	• Market incentives	$10.0
	• Maternal and child health interventions and service delivery	$17.0
	• Support for international organizations (PAHO, UNICEF)	$13.0
USAID Total		*$137.0*
CDC	• Implement Vector Management and Control	$7.0
	• Maternal and Child Health Interventions	$1.0
	• Innovations	$5.0
	• Surveillance, Epidemiology, and Studies	$44.0
	• Laboratory Capacity and Equipment	$15.0
	• Management and Coordination of Response	$6.0
CDC Total		*$78.0*

Source: USAID Congressional Notification, "Zika Response: Initial Time-Critical Activities," April 6, 2016.

APPENDIX B. ONLINE RESOURCES ON ZIKA VIRUS

Organization	Source Title	Document/Link
Centersfor Disease Control and Prevention (CDC)	Zika Travel Information	http://wwwnc.cdc.gov/travel/page/zika-travel-information
CDC	Zika Virus Index	http://www.cdc.gov/zika/index.html
Pan American Health Organization (PAHO)	Zika Resources for Health Authorities	http://www.paho.org/hq/index.php?option=com_content&view=article&id=11601&Itemid=41694&lang=en
PAHO	Zika Resources for General Public	http://www.paho.org/hq/index.php?option=com_content&view=article&id=11602&Itemid=41695&lang=en
PAHO	Epidemiological Alerts and Updates	http://www.paho.org/hq/index.php?option=com_content&view=article&id=1239&Itemid=2291&lang=en
World Bank	Zika Virus Fact Sheet	http://www.worldbank.org/en/topic/health/brief/world-bank-group-and-zika-fact-sheet
UNICEF, WHO, PAHO, IFRC	Risk communication and community engagement for Zikavirus prevention and control	http://www.unicef.org/cbsc/files/Zika_Virus_Prevention_and_Control_UNICEF_English.pdf
United Nations Population Fund (UNFPA)	Zika virus: Frequently asked questions	http://www.unfpa.org/resources/zika-virus-frequently-asked-questions
UNFPA	UNFPA on the Zika Virus Outbreak	http://www.unfpa.org/press/unfpa-zika-virus-outbreak
World Health Organization (WHO)	Zika Strategic Response Framework and Joint Operations Plan	http://apps.who.int/iris/bitstream/10665/204420/1/ZikaResponseFramework_JanJun16_eng.pdf?ua=1
WHO	Zika Virus Situation Reports	http://who.int/emergencies/zika-virus/situation-report/en/

End Notes

[1] WHO, Situation Report: Zika Virus, Microcephaly, Guillain-Barré Syndrome, June 16, 2016.

[2] Ibid

[3] WHO, "WHO Statement on the First Meeting of the International Health Regulations (2005) Emergency Committee on Zika Virus and Observed Increase in Neurological Disorders and Neonatal Malformations," February 1, 2016.

[4] PAHO is the WHO's regional office for the Americas.

[5] Local transmission means that mosquitoes in the area have been infected with the virus and are spreading it to people.

[6] WHO, Prevention of Sexual Transmission of Zika Virus: Interim Guidance Update, June 7, 2016.

[7] As an example, children from low-income families are five times as likely to die before the age of five than children from wealthier households. World Bank, "Latin America: Unequal Access to Health Care Is Still No. 1 Killer for Moms and Kids," September 11, 2013.

[8] CARICOM is a grouping of 20 countries: 15 member states and five associate members. The 15 member states include Antigua and Barbuda, Bahamas, Barbados, Belize, Dominica, Grenada, Guyana, Haiti, Jamaica, Montserrat, Saint Lucia, St. Kitts and Nevis, St. Vincent and the Grenadines, Suriname, and Trinidad and Tobago. The five associate member states include Anguilla, Bermuda, the British Virgin Islands, the Cayman Islands, and Turks and Caicos. CARICOM's four main purposes are to (1) promote economic integration and cooperation among the community, (2) facilitate foreign policy coordination, (3) advance human and social development, and (4) ensure mutual security. For more information, see http://caricom.org.

[9] For background on GHSA, see CRS In Focus IF10022, The Global Health Security Agenda and International Health Regulations, by Tiaji Salaam-Blyther.

[10] WHO, Zika Virus, April 15, 2016.

[11] M. K. Kindhauser et al., "Zika: The Origin and Spread of a Mosquito-Borne Virus," Bulletin of the World Health Organization, February 9, 2016.

[12] Lulan Wang et al., "From Mosquitoes to Humans: Genetic Evolution of Zika Virus," Cell Host and Microbe, vol. 19 (May 11, 2016).

[13] Kindhauser, "Zika.". GBS is a condition in which a person's immune system attacks the peripheral nerves. Many people who develop GBS recover fully, including those with severe GBS. Severe cases of GBS are rare but can result in death. For more information on GBS, see WHO, Guillain-Barré Syndrome, March 14, 2016.

[14] Microcephaly is a condition in which a baby is born with a small head or the head stops growing after birth. Some babies born with mild microcephaly can live normal lives, while most babies born with severe microcephaly can experience epilepsy, cerebral palsy, learning disabilities, hearing loss, and vision problems over their lifetimes. See WHO, Microcephaly, March 2, 2016.

[15] WHO, Situation Report.

[16] According to the World Bank, at least 110 million people in Latin America lack access to modern sanitation. World Bank, "Monitoring Country Progress in Water and Sanitation," June 13, 2014.

[17] CRS In Focus IF10353, Mosquitoes, Zika Virus, and Transmission Ecology, by M. Lynne Corn, Tadlock Cowan, and Robert Esworthy.

[18] WHO, "El Niño May Increase Breeding Grounds for Mosquitoes Spreading Zika Virus, WHO Says," February 22, 2016.

[19] Sonia Altizer et al., "Climate Change and Infectious Diseases: From Evidence to a Predictive Framework," Science, vo. 341 (August 15, 2013), p. 514.

[20] Chikungunya is a virus that can cause fever and severe joint pain, which can be debilitating. Other symptoms typically include nausea, fatigue, rash, and muscle pain. There is no

antiviral treatment or vaccine. From 2013 through 2015, more than 1.7 million people contracted the disease, and 258 people died. During that time period, CDC estimated that 3,113 cases were imported into the United States and 11 cases were contracted locally in Florida. From the beginning of 2016 through June 17, 2016, PAHO/WHO reported 130,138 chikungunya cases in the region, of which 18,220 have been confirmed. On May 31, 2016, the Texas Department of State Health Services reported the first locally acquired chikungunya case, indicating that mosquitoes within the United States now carry the virus. WHO, "Fact Sheet: Chikungunya," April 2016; U.S. Department of Defense, Chikungunya in the Americas Surveillance Summary #49, September 9, 2015; WHO/PAHO, Number of Reported Cases of Chikungunya Fever in the Americas, by Country or Territory 2016, June 17, 2016; Texas Department of State Health Service, "DSHS Announces First Texas Acquired Chikungunya Case," press release, May 31, 2016.

21 Dengue is a virus that can cause severe, flu-like symptoms and can cause death in about 1%-2% of all cases. Severe dengue is evidenced by severe abdominal pain, repeated vomiting, rapid breathing, bleeding gums, or blood in vomit. It is a leading cause of death for children in some Latin American countries. There is no specific treatment for dengue, but a vaccine was introduced in a few countries in late 2015. From the beginning of 2014 through June 17, 2016, PAHO estimated that more than 5 million people had contracted dengue in the region and more than 2,500 people died of the disease. Some 1,299 of these cases occurred in the United States, and no deaths were reported. WHO, "Fact Sheet: Dengue and Severe Dengue," April 2016 and WHO/PAHO, Number of Reported Cases of Chikungunya Fever in the Americas.

22 Greg Botelho, "Zika Virus 'Spreading Explosively,' WHO Leader Says," CNN, February 20, 2016.

23 PAHO/WHO, Cumulative Zika Suspected and Confirmed Cases Reported by Countries and Territories in the Americas, 2015-2016, June 16, 2016.

24 Ibid.

25 CDC, "Interim Guidance for Zika Virus Testing of Urine—United States, 2016," Morbidity and Mortality Weekly Report, vol. 65, no. 18 (May 13, 2016).

26 CDC, "Zika Virus Disease in the United States, 2015-2016," http://www.cdc.gov/zika/geo/united-states.html (accessed on June 16, 2016).

27 Dan Diamond, "Frieden: CDC Will Lose Zika Fight Without Funding," Politico, May 26, 2016.

28 This text box was authored by Sarah A. Lister, Specialist in Public Health and Epidemiology.

29 CRS correspondence with health experts at the Inter-American Development Bank, June 22, 2016.

30 PAHO, Strategic Plan of the Pan-American Health Organization: 2014-2019, September 2013. As of 2012, PAHO deemed six countries—Brazil, Canada, Chile, Colombia, Costa Rica, and the United States—capable of addressing pandemic outbreaks.

31 Guttmacher Institute, "Fact Sheet: Abortion Laws in Latin America and the Caribbean," May 2016.

32 National Public Radio, "CDC Waits for Congress to Approve Emergency Funds to Combat Zika," May 17, 2016.

33 Peter J. Meyer, Analyst in Latin American Affairs, contributed to this section. For more on Brazil, see CRS Report RL33456, Brazil: Background and U.S. Relations, by Peter J. Meyer.

34 WHO, Situation Report.

35 Ibid.

[36] Marcia Castro, "Zika Virus and Health Systems in Brazil: From Unknown to a Menace," Health Systems and Reform, May 3, 2016.

[37] Brent McDonald, "Brazil's Abortion Restrictions Compound Challenge of Zika Virus," New York Times, May 18, 2016.

[38] Luiz Alberto Figueiredo Machado, "Brazil Is Doing Its Part in the Global Fight Against Zika," Americas Quarterly, February 4, 2016.

[39] Kwame A. Nyarko et al., "Explaining Racial Disparities in Infant Health in Brazil," American Journal of Public Health, 2013, p. 103; Olga Khazan, "What the U.S. Can Learn from Brazil's Healthcare Mess," The Atlantic, May 8, 2014.

[40] Castro, 2016; Stephen Eisenhammer and Pedro Fonseca, "Brazil Health Service Cracking Under Strain of Microcephaly," Reuters, February 23, 2016.

[41] WHO/PAHO, Cumulative Zika Suspected and Confirmed Cases.

[42] CRS communication with the Ministry of Health in Colombia, May 23, 2016.

[43] Ibid.

[44] WHO, Situation Report.

[45] Colombian law reportedly allows abortions for pregnancies resulting in "malformations incompatible with life," as well as those that pose a "risk the physical, mental, or social health" of a woman. See Justin Calderon, "Colombia's Pro-Lifers Are Objectively Pro-Zika," Foreign Policy, April 18, 2015.

[46] CRS communication with the CDC on May 19, 2016; El Colombiano, "En Colombia Hay Cinco Casos Confirmados de Microcefalia por Zika," May 21, 2016.

[47] WHO, "WHO to Convene an International Health Regulations Emergency Committee on Zika Virus and Observed Increase in Neurological Disorders and Neonatal Malformations," January 28, 2016.

[48] WHO, Zika Virus Outbreak Global Response: Interim Report, May 27, 2016.

[49] UNICEF is working with governments and local communities to prevent the spread of Zika by eliminating mosquito breeding sites and distributing insect repellant and bed nets. UNICEF is also providing support for the care of infants born with microcephaly. The U.N. Population Fund is helping women of childbearing age in Zika-affected countries access information about Zika in pregnancy, contraception (to reduce the risk of contracting Zika while pregnant), and counseling services.

[50] WHO, Zika: Response Funding, June 22, 2016, http://who.int/emergencies/zika-virus/response/contribution/en.

[51] WHO, Zika: Strategic Response Plan, June 2016.

[52] Those dimensions include the country's capacity to (1) conduct surveillance of the disease, (2) diagnose cases, (3) launch an emergency response, (4) coordinate among various ministries, (5) conduct research, (6) monitor and report on arboviruses, (7) monitor and control complications related to Zika, (8) manage costs, and (9) provide needed health services.

[53] CRS phone interview with World Bank officials, May 12, 2016.

[54] Worid Bank, The Short-Term Costs of Zika on Latin America and the Caribbean, February 18, 2016.

[55] Center for Strategic and International Studies (CSIS), "2016 Global Development Forum: Combating Infectious Disease: The Unfolding Threat of Zika," May 19, 2016.

[56] The International Development Assistance countries are Bolivia, Dominica, St. Vincent, Grenada, Guyana, Haiti, Honduras, St. Lucia, and Nicaragua.

[57] Information in the remainder of this section is from CRS communication with World Bank officials, June 16, 2016.

[58] The International Health Regulations (2005) requires WHO Member States to (1) notify WHO of any event that may constitute a PHEIC and respond to requests for verification of information regarding such events, (2) follow WHO recommendations concerning appropriate public health responses to the relevant PHEIC, (3) build and maintain core public health capacities for disease surveillance and response, and (4) collaborate with other member states to provide or facilitate the delivery of technical assistance in support of developing and maintaining core public health capacities among all member states. See CRS In Focus IF10022, The Global Health Security Agenda and International Health Regulations, by Tiaji Salaam-Blyther.

[59] CRS correspondence with IDB, May 17, 2016.

[60] CRS correspondence with USAID, June 22, 2016.

[61] CRS interview with USAID personnel, May 3, 2016.

[62] This section draws from CRS Report R44460, Zika Response Funding: Request and Congressional Action, coordinated by Susan B. Epstein.

[63] CRS Report R44460, Zika Response Funding: Request and Congressional Action, coordinated by Susan B. Epstein.

[64] CRS correspondence with PAHO official, June 10, 2016.

[65] NIH, "PAHO's Dr. Marcos Espinal on Global Partnerships to Fight Zika, Ebola, and Other Diseases," Global Health Matters, April 2016; CSIS, "2016 Global Development Forum."

[66] That estimate includes funds needed for a joint response by WH0/PAHO and some 60 partners. WHO, Zika Strategic Response Plan Revised for July 2016-December 2017, June 2016.

[67] USAID, "Zika Response: Initial Time-Critical Activities," April 6, 2016.

[68] For more information, see CRS Report R44113, U.S. Foreign Assistance to Latin America and the Caribbean: Recent Trends and FY2016 Appropriations, by Peter J. Meyer.

[69] See White House, "U.S. Global Development Policy," September 22, 2010.

[70] CRS correspondence with USAID, June 8, 2016.

[71] Jane T. Bertrand, USAID Graduation From Family Planning Assistance: Implications for Latin America, Population Institute and Tulane University School of Public Health and Tropical Medicine, October 2011; correspondence with USAID, June 8, 2016.

[72] U.N. Development Program, Multidimensional Progress: Well-Being Beyond Income, June 2016

[73] Kimberly Cole, "A USAID Legacy in Latin America: Smaller Families and Better Health," Frontlines, July/August 2013.

[74] Excludes regional programs in Latin America and the Caribbean that focus on HIV/AIDS.

[75] Sarah Bott et al. Violence Against Women in Latin America and the Caribbean: A Comparative Analysis of Population-Based Data from 12 Countries, PAHO, 2013.

[76] G. Sedgh, S. Singh, and R. Hussain, "Intended and Unintended Pregnancies Worldwide in 2012 and Recent Trends," Studies in Family Planning, no. 3 (September 2014), pp. 301-14.

[77] Margaret Chan, address to the 69th World Health Assembly, Geneva, Switzerland, May 23, 2016. See also Woodrow Wilson Center, "How Zika Is Shaping the Sexual and Reproductive Health and Rights Agenda," April 12, 2015.

[78] For current restrictions on U.S. family planning assistance, see CRS Report R41360, Abortion and Family Planning-Related Provisions in U.S. Foreign Assistance Law and Policy, by Luisa Blanchfield.

[79] For more information on GHSA, see CRS In Focus IF10022, The Global Health Security Agenda and International Health Regulations, by Tiaji Salaam-Blyther.

[80] Victor J. Dzau and Peter Sands, "Beyond the Ebola Battle—Winning the War Against Future Epidemics," New England Journal of Medicine, June 2016; Lawrence O. Goslin,

"Neglected Dimensions of Global Security: The Global Health Risk Framework Commission," Journal of the American Medical Association, vol. 315, no. 12 (April 12, 2016).

[81] PAHO, September 2013.

[82] See the GHSA website at http://www.cdc.gov/globalhealth/security/ghsagenda.htm.

[83] See, for example, NIH, "NIAID Research Approach to Zika Virus," https://www.niaid.nih.gov/topics/zika/ ResearchApproach/Pages/default.aspx; and DOD, "Walter Reed Scientists Test Zika Vaccine Candidate," June 10, 2016.

[84] WHO, Investing to Overcome the Global Impact of Neglected Tropical Diseases: Third WHO Report on Neglected Tropical Diseases, February 2015; and Helen M. Lazear, Elizabeth M. Stringer, and Aravinda M. de Silva, "The Emerging Zika Virus Epidemic in the Americas: Research Priorities," Journal of the American Medical Association, May 10, 2016.

[85] Global Health Technologies Coalition, Achieving a Bold Vision for Global Health: Policy Solutions to Advance Global Health R&D, 2016, p. 6.

[86] Ibid.

[87] Statement of Sophie Delaunay, Adviser, Doctors Without Borders, Committee on Senate Foreign Relations, Subcommittee on Africa and Global Health Policy, April 7, 2016.

[88] Margaret Chan, "Address to the Sixty-Ninth World Health Assembly," May 23, 2016.

[89] Ildefonso Fernandez-Salas, "Historical Inability to Control Aedes Aegypti as a Main Contributor of Fast Dispersal of Chikungunya Outbreaks in Latin America," Antiviral Research, vol. 124 (October 27, 2015).

[90] For more information on genetically modified mosquitoes, see CRS In Focus IF10401, Genetically Engineered Mosquitoes: A Vector Control Technology for Reducing Zika Virus Transmission, by Tadlock Cowan.

[91] Kelsey Snell, "Zika Funding Bill Blocked in the Senate," Washington Post, June 28, 2016.

In: The Zika Virus
Editor: Veronica Salazar

ISBN: 978-1-53610-124-9

Chapter 3

ZIKA RESPONSE FUNDING: REQUEST AND CONGRESSIONAL ACTION*

Susan B. Epstein and Sarah A. Lister

SUMMARY

The second session of the 114th Congress is considering whether and how to provide funds to control the spread of the Zika virus throughout the Americas. Zika infection, primarily spread by *Aedes* mosquitoes, has been linked to severe birth defects and other health concerns. Local transmission of the Zika virus has occurred in American Samoa, Puerto Rico, and the U.S. Virgin Islands, and is expected on the U.S. mainland this summer, in areas where *Aedes* mosquitoes are present.

On February 22, 2016, the Obama Administration submitted a request for more than $1.89 billion in supplemental funding to respond to the Zika epidemic, all of which is requested as emergency discretionary appropriations and therefore effectively exempt from spending limits per the Budget Control Act of 2011 (BCA, P.L. 112-25). The emergency request includes $1.509 billion for the Department of Health and Human Services (HHS), $335 million for the U.S. Agency for International Development (USAID), and $41 million for the Department of State. The request also seeks authority to transfer some of those supplemental emergency appropriations across other

* This is an edited, reformatted and augmented version of a Congressional Research Service publication R44460, prepared for Members and Committees of Congress, dated June 2, 2016.

federal agencies such as the Department of Defense, the Environmental Protection Agency, and the U.S. Department of Agriculture, to allow greater flexibility as circumstances change. It also would provide HHS, the Department of State, and USAID with broad authority for direct hiring, not be limited to positions related to Zika response efforts, and would authorize personal services contracting by HHS, State and USAID, for the Zika virus, but not limited to Zika virus efforts.

On April 6, 2016, the White House Office of Management and Budget (OMB) and the Secretary of HHS announced that they had identified $589 million—$510 million of it from "existing Ebola resources within the Department of Health and Human Services and Department of State/USAID"—that can quickly be redirected and spent on immediate efforts to control and respond to the spread of the Zika virus in the Americas.

On April 8, 2016, the Administration notified Congress of a transfer of $295 million out of the $510 million from FY2015 unobligated Ebola Economic Support Funds (ESF) to be used for the Zika response efforts. Of that amount, USAID is providing $158 million to CDC including $78 million for Zika response and $80 million for Ebola response. The remaining $137 million also from FY2015 ESF is funding various USAID activities for its Zika response efforts.

As of mid-May 2016, congressional action on supplemental appropriations for Zika-related purposes has occurred in both the House and the Senate. The Senate action occurred with regard to an amendment to the combined FY2017 Military Construction-Veterans Affairs and Transportation-Housing and Urban Development appropriations bills (S.Amdt. 3900); that amendment provides $1.1 billion, which would be available until September 30, 2017 (some until expended). House action has occurred with regard to a stand-alone supplemental appropriations bill (H.R. 5243); this bill provides $622.1 million for Zika funding, which would be available until September 30, 2016. Nearly half of the funds are designated as emergency funding. Rescissions are included. CBO scored the total budgetary effects of these rescissions as reducing budget authority by $622 million, and reducing outlays by $12 million during FY2016. On net, CBO estimates the budgetary effects of the bill to be $0 in budget authority and an increase of $9 million in outlays for FY2016.

This report will identify the various Zika response funding options and track legislation in the 114th Congress.

INTRODUCTION

In its second session, the 114th Congress is considering whether and how to provide funds to control the spread of the Zika virus throughout the

Americas. Zika infection, primarily spread by *Aedes* mosquitoes, has been linked to severe birth defects and other health concerns. Local transmission of the Zika virus has occurred in American Samoa, Puerto Rico, and the U.S. Virgin Islands, and transmission is expected on the U.S. mainland this summer, in areas where *Aedes* mosquitoes are present.[1]

Federal efforts to address the outbreak include research on the infection and its effects, mosquito control measures, and efforts to develop a vaccine. The public health focus, both domestically and elsewhere in the Americas, is to protect pregnant women from infection and prevent birth defects. Administration officials and some in Congress are concerned about the resources needed to prevent widespread Zika infections as the Northern Hemisphere summer approaches.

This report presents the Administration's request for supplemental appropriations for the Zika response, supplemental appropriations measures that have received congressional action, and information about unobligated Ebola supplemental funds as of January 1, 2016 (the most recent publicly available).

Administration Actions

On February 22, 2016, the Obama Administration submitted a request for more than $1.89 billion in supplemental funding to respond to the Zika epidemic, all of which is requested as emergency FY2016 discretionary appropriations, and therefore effectively exempt from spending limits in the Budget Control Act of 2011 (BCA, P.L. 112-25).[2] The Administration's request includes $1.509 billion for the Department of Health and Human Services (HHS), $335 million for the U.S. Agency for International Development (USAID), and $41 million for the Department of State. The request also seeks authority to transfer some of those supplemental emergency appropriations across other federal agencies, such as the Department of Defense, the Environmental Protection Agency, and the U.S. Department of Agriculture, to allow greater flexibility as circumstances change. It also would provide HHS, the Department of State, and USAID with two personnel management authorities for, but not limited to, addressing the Zika outbreak; broad authority for direct hiring;[3] and authority for personal services contracting.[4]

A major theme in congressional debates thus far has been over whether unobligated (uncommitted for expenditure) FY2015 funds that had been provided to respond to the Ebola virus outbreak should be used to fund part of the Zika response, either temporarily or permanently.[5] On April 6, 2016, the White House Office of Management and Budget (OMB) and the Secretary of HHS announced that they had identified \$589 million—\$510 million of it from "existing Ebola resources within the Department of Health and Human Services and Department of State/USAID"—that could quickly be reprogrammed and spent on immediate efforts to control and respond to the spread of the Zika virus in the Americas.[6]

As part of the reprogramming, on April 8, 2016, USAID notified Congress of its intent to redirect \$295 million of the \$510 million from FY2015 unobligated Ebola Economic Support Funds (ESF) to be used for Zika response. Of that amount, USAID transferred \$158 million to CDC, including \$78 million for Zika response and \$80 million for Ebola response. The remaining \$137 million, also from FY2015 ESF, is to be redirected to fund various USAID activities for Zika response efforts.

CONGRESSIONAL ACTIONS

On May 16, 2016, the Chairman of the House Appropriations Committee introduced the Zika Response Appropriations Act, 2016 (H.R. 5243). The bill would provide \$622.1 million that would be available until September 30, 2016, for domestic and international Zika response efforts. Nearly half of the funds are designated as emergency funds. The bill also includes rescissions of certain Ebola-related appropriations and the HHS nonrecurring expenses fund.[7] Provisions in the bill provide that its appropriations shall be subject to the same requirements for funds that applied to the Consolidated Appropriations Act, 2016 (P.L. 114-113). This would include any restrictions on the use of funds that were contained therein, such as the applicable prohibitions on the use of funds for abortions. The House passed the measure on May 18, 2016, without amendment.

Senate action has occurred in the context of the FY2017 Military Construction-Veterans Affairs and Transportation-Housing and Urban Development appropriations bills. The texts of these bills were combined for the purposes of initial consideration in the Senate, and offered as a substitute amendment to an unrelated measure (S.Amdt. 3896 to H.R. 2577). On May 12, 2016, Senator McConnell (on behalf of Senator Blunt) proposed an amendment to that substitute (S.Amdt. 3900 to S.Amdt. 3896) that would provide \$1.1 billion for Zika response and preparedness, which would be available until

September 30, 2017 (with the exception of Global Health Funds, which would be available until expended). On May 19, 2016, the Senate adopted S.Amdt. 3900,[8] followed by the pair of appropriations measures (i.e., H.R. 2577) to which it was amended.

See *Table 1* below for a comparison of the Administration's supplemental request, S.Amdt. 3900, and House bill H.R. 5243. For a comparison of non-monetary provisions of these measures, see *Table 2*.

Table 1. Supplemental Funding Amounts for Zika Response for FY2016: Comparison of Administration Request with Senate and House Proposals

Budget Authority in $ Millions

Agency	Administration Request	S.Amdt. 3900 as Passed in Senate	H.R. 5243 as Passed in House
CDC Subtotal	828.0	449.0[a]	170.0[b,c]
HRSA: Community Health Centers for territories (non-add)	*0.0*	*(40.0)*	*0.0*
HRSA: National Health Service Corps for territories (non-add)	*0.*[0d]	*(6.0)*	*0.0*
HRSA: Maternal and Child Health Block Grant (non-add)	*0.0*[d]	*(5.0)*	*0.0c*
HRSA Subtotal	0.0	51.0	0.0
PHSSEF: Social Services Block Grant for territories (non-add)	*0.0*	*(75.0)*	*0.0*
PHSSEF: Other (non-add)	*(295.0)*[d]	*(75.0)*[e]	*(103.0)*
PHSSEF Subtotal	295.0	150.0	103.0
NIH/NIAID Subtotal	130.0	200.0	230.0
FDA Subtotal	10.0	0.0	0.0[f]
CMS Subtotal (Medicaid federal matching rate)	246.0	no provision	no provision
HHS Total	1,509.0	850.0[g]	503.0
State: Diplomatic and Consular Programs (D&CP)	14.6	14.6	9.1[h]
State: Emergencies in the Diplomatic and Consular Service	4.0	4.0	0.0[i]
State: Repatriation Loans	1.0	1.0	0.0
State: Nonproliferation, Anti-Terrorism, Demining and Related Programs (NADR)	8.0	4.0	0.0
State: International Organizations and Programs (IO&P)	13.5	13.5	0.0
USAID: Operating Expenses (OE)	10.0	10.0	10.0

Table 1. (Continued)

Agency	Administration Request	S.Amdt. 3900 as Passed in Senate	H.R. 5243 as Passed in House
USAID: Global Health Programs (GHP)	325.0	211.0	100.0[j]
State / USAID Total	376.1	258.1	119.1
REQUEST OR BILL TOTAL	1,885.1	1,108.1	622.1

Source: CRS analysis of text of White House, Office of Management and Budget, "Estimate #1–FY 2016 Emergency Supplemental: Appropriations Request to Respond to the Zika Virus both Domestically and Internationally," February 22, 2016, https://www.whitehouse.gov/omb/budget_amendments; S.Amdt. 3900; and H.R. 5243 IH.

Notes: Numbers in parentheses are included in subtotals.

[a] Of the CDC funds provided, $88 million may be used to reimburse prior Zika response spending.

[b] Of the CDC funds provided, up to $500,000 each must be transferred to the HHS Office of Inspector General and the Comptroller General for oversight activities.

[c] Up to $50.0 million of the CDC funds provided may be transferred to HRSA MCH for specified activities.

[d] Unspecified amounts from the PHSSEF may be transferred to HRSA for National Health Service Corps activities in the territories, and for the Maternal and Child Health Block Grant.

[e] Other Public Health and Social Services Emergency Fund (PHSSEF) activities principally involve development and purchase of medical countermeasures.

[f] The House-reported Agriculture and Related Agencies appropriation for FY2017 included $10 million for FDA activities related to the response to Ebola, Zika, and other emerging threats. H.Rept. 114-531, p. 70.

[g] Broad authority is provided for funds to be transferred to or merged with other CDC, PHSSEF, HRSA, and NIH appropriations accounts to fund the purposes specified in the measure. Such transfers can only occur following consultation with OMB with a 10-day notification in advance, among other requirements.

[h] Up to $1.35 million of funds for Diplomatic and Consular Services may be used for medical evacuation costs for any U.S. agency.

[i] Up to $1.0 million of funds for Diplomatic and Consular Services may be transferred to Emergencies in the Diplomatic and Consular Service

[j] Of the Global Health Program funds provided, up to $500,000 each must be transferred to the USAID Office of Inspector General and the Comptroller General for oversight activities.

Table 2. Selected Supplemental Provisions for Zika Response for FY2016: Comparison of Administration Request with Senate and House Proposals

Provision	AdministrationRequest	Senate(S.Amdt. 3900)	House(H.R. 5243IH)
Period of Availability of Funds			
HHS Funds	Until expended.	Until Sept. 30, 2017.	Until Sept. 30, 2016.
State / USAID Funds: D&CP:Nonproliferation, Anti-Terrorism, Demining and Related Programs;International Organizations and Programs;Operating Expenses	Until Sept. 30, 2017.	Until Sept. 30, 2017.	Until Sept. 30, 2016.
State / USAID Funds: Emergencies in the Diplomatic and Consular Service;RepatriationLoans;Global Health Programs.	Until expended.	Until expended.	Until Sept. 30, 2016.
Scope of Use of Funds			
CDC Funds	To prevent, prepare for, and respond to Zikavirus, other vector-borne diseases, or other infectious diseases and related health outcomes, domestically and internationally.	To prevent, prepare for, and respond to Zika virus, other vector-borne diseases, and related health outcomes, domestically and internationally.	To prevent, prepare for, andrespond to Zika virus, domestically and internationally.
HRSA Funds	(Scope for PHSSEF funds would apply to any funds transferred to HRSA.)	To prevent, prepare for, and respond to Zika virus, other vector-bornediseases, and related health outcomes, domestically and internationally.	(Scope for CDC funds wouldapply to any funds transferred to HRSA.)

Table 2. (Continued)

Provision	AdministrationRequest	Senate(S.Amdt. 3900)	House(H.R. 5243IH)
NIH Funds	To prevent, prepare for, and respond to Zika virus, other vector-borne diseases, or other infectious diseases and related health outcomes, domestically and internationally.	To prevent, prepare for, and respond to Zika virus, other vector-borne diseases, and related health outcomes, domestically and internationally.	For development of vaccines for the Zikavirus.
PHSSEF Funds	To prevent, prepare for, and respond to Zika virus, other vector-borne diseases, or other infectious diseases and related health outcomes, domestically and internationally.	To prevent, prepare for, and respond to Zika virus, other vector-borne diseases, and related health outcomes, domestically and internationally.	To respond to Zika virus, domestically and internationally.
State Dept. Diplomatic and Consular Programs (D&CP)	To support response efforts related to the Zika virus and related health outcomes, other vector-borne diseases, or other infectious diseases.	To support response efforts related to the Zika virus and related health outcomes, other vector-borne diseases, or other infectious diseases.	To support cost of medical evacuations and other response efforts related to the Zika virus and health conditions directly associatedwith the Zika virus.
State Dept. Emergencies in Diplomatic and Consular Service	To support response efforts related to the Zika virus and related health outcomes, other vector-borne diseases.	To support response efforts related to the Zika virus and related health outcomes, other vector-borne diseases.	No comparable provision.
Repatriation Loans Program	For direct loans to support response efforts related to the Zika virus and related health outcomes, other vector-borne diseases, or other	For direct loans to support response efforts related to the Zika virus and related health outcomes, other	No comparable provision.

Provision	AdministrationRequest	Senate(S.Amdt. 3900)	House(H.R. 5243IH)
	infectious diseases.	vector-borne diseases, or other infectious diseases.	
USAID Operating Expenses (OE)	To support response efforts related to the Zika virus and related health outcomes, other vector-borne diseases, or other infectious diseases.	To support response efforts related to the Zika virus and related health outcomes, other vector-borne diseases, or other infectious diseases.	Response efforts related to the Zika virus and health conditions directly associatedwith the Zika virus.
Global Health Programs (GHP)	For assistance or research to prevent, treat, or otherwise respond to the Zika virus and related health outcomes, other vector-borne diseases, or other infectious diseases.	For assistance or research to prevent, treat, or otherwise respond to the Zika virus and related health outcomes, other vector-borne diseases, or other infectious diseases.	For vector control activities to prevent, prepare for, and respond to the Zika virus internationally.
Dept. of State, Nonproliferation, Anti-terrorism, Demining and Related Programs (NADR)	To support response and research efforts related to the Zika virus and related health outcomes, other vector-borne diseases, or other infectious diseases.	To support response and research efforts related to the Zika virus and related health outcomes, other vector-borne diseases, or other infectious diseases.	No comparable provision.
International Organizations and Programs (IO&P)	To support response and research efforts related to the Zika virus and related health outcomes, other vector-borne diseases, or other infectious diseases.	To support response and research efforts related to the Zika virus and related health outcomes, other vector-borne diseases, or other infectious	No comparable provision.

Table 2. (Continued)

Provision	AdministrationRequest	Senate(S.Amdt. 3900)	House(H.R. 5243IH)
		diseases.	
Transfer Authority			
HHS Funds	CDC funds may be transferred within CDC. NIH funds may be transferred within NIH. PHSSEF funds may be transferred to twostated HRSA accounts, as specified, to an HHS countermeasures injury compensation fund, and to any other HHS accounts.	Any HHS funds in the amendment may be transferred to accounts in CDC, HRSA, NIH, and PHSSEF. $75 million in PHSSEF funds must be transferred to the HHS Social Services Block Grant.	CDC funds may be transferred within CDC, and to three stated HRSA accounts, as specified. NIH funds may be transferred within NIH. PHSSEF funds may be transferred to an HHS countermeasures injury compensation fund.
International Affairs	Funds may be transferred between foreign affairs accounts within the same headings to carry out the purposes of this Act and are in addition to other transfer authority within this proposal.	Funds within certain foreign affairs accounts may be transferred between foreign affairs accounts within the same headings to carry out the purposes of this Act and are in addition to other transfer authority within this proposal.	Specified funds within D&CP may be transferred for medical evacuation, transferred for Emergencies in Diplomatic and Consular Services, and are in addition to any other transfer authority within this proposal.
Notification,Reportingand Oversight			
HHS Notification Requirement for Obligation	No comparable provision.	No comparable provision.	15 days in advance of obligation.
International Affairs Notification Requirement	No comparable provision.	15 days in advance of obligation.	15 days in advance of obligation.
HHS Reporting Requirement	No comparable provision.	Within 30 days of enactment the	Within 30 days of enactmentthe HHS

Provision	AdministrationRequest	Senate(S.Amdt. 3900)	House(H.R. 5243IH)
		HHS Secretary must report to the Appropriations Committees with a spend plan, followed by quarterly reports on obligations until funds have been fully expended.	Secretary must report to the AppropriationsCommittees with a spend plan, which must be updated and resubmitted every 30 days until funds have been fully expended.
International Affairs Reporting Requirement	No comparable provision.	Within 45 days after enactment and prior to obligation of international funds, the USAID	Within 30 days after enactment the Secretary of State and USAID Administrator must submit
		Administrator must submit spend plans to the Committees on Appropriations, update and resubmit to those committees every 90 days until September 30, 2017, and every 180 days thereafter until all funds are expended.	to Appropriations Committees a consolidated report and update and submitted to those committees every 30 days until all funds are expended.
HHS Oversight	No comparable provision.	No comparable provision.	$500,000 of CDC funds mustbe made available to the HHSOffice of the Inspector General for oversight; an additional $500,000 from CDCmust bc made available for

Table 2. (Continued)

Provision	AdministrationRequest	Senate(S.Amdt. 3900)	House(H.R. 5243IH)
			oversight by the Comptroller General of the United States.
International Affairs Oversight	No comparable provision.	$500,000 made available within the International Affairs Chapter isto be made available for the Comptroller General of the United States for oversight activities and for consultation with the Secretary of State and the USAID Administrator prior to obligating such funds.	$500,000 from GHP within this Title must be made available to USAID's Office of the Inspector General for oversight; an additional $500,000 from GHP must be made available for oversight by the Comptroller General of the United States.

Source: CRS analysis of text of White House, Office of Management and Budget, "Estimate #1–FY 2016 Emergency Supplemental: Appropriations Request to Respond to the Zika Virus both Domestically and Internationally," February 22, 2016, https://www.whitehouse.gov/omb/budget_amendments; S.Amdt. 3900; and H.R. 5243 IH.

The Emergency Supplemental Appropriations Request for Zika Response Efforts

The following describes the Administration's February 2016 Zika emergency supplemental request components by agency.

It was reported on April 18 that the Administration submitted a revised Zika supplemental request to Congress, which would maintain departmental request totals, while redirecting some of the HHS funds requested for contingency use to vaccine research and development at the National Institutes of Health (NIH).[9] Detailed information about this is not publicly available, and the following narrative does not reflect this revision.

Health and Human Services

The Administration's emergency supplemental appropriations request to respond to the Zika outbreak seeks for HHS a total of $1.509 billion. Each HHS agency request includes the statement that funds would be "to prevent, prepare for, and respond to Zika virus, other vector-borne diseases, or other infectious diseases and related health outcomes, domestically and internationally.... " Most of the requested funds would support research, surveillance, vaccine and test development, and various domestic preparedness activities. A portion would support international response activities. The request proposes that all supplemental appropriations to HHS be designated as emergency spending, and remain available until expended.

Centers for Disease Control and Prevention (CDC)

A total of $828 million of the February 2016 request is for the CDC-Wide Activities and Program Support account. Proposed request language would, among other things, authorize the CDC Director to transfer funds between CDC accounts, and authorize funds to be used for real property acquisition and improvements to non-federal facilities. Funds would be used as follows:

- *Grants and technical assistance to Puerto Rico and U.S. Territories*—$225 million to, among other purposes, monitor pregnant women and establish a registry of women infected while pregnant; expand mosquito control activities; and enhance laboratory testing capacity.
- *Domestic Response*—$453 million to provide grants to southern and other U.S. states with *Aedes* mosquitoes for surveillance, improved test methods and testing capacity, public education and outreach, mosquito control measures in areas at risk, and additional federal and state response activities.
- *International Response Activities*—$150 million to expand the public health workforce, and enhance infectious disease surveillance and emergency response activities, in Zika-affected countries; and to support the laboratory network of the Pan American Health Organization (PAHO), the regional arm of the World Health Organization (WHO) for the Americas.

Public Health and Social Services Emergency Fund (PHSSEF)

The PHSSEF is a fund used by appropriators to provide the HHS Secretary with ongoing or onetime emergency funding, such as for the response to disease epidemics. The emergency supplemental request seeks $295 million for the PHSSEF for the following:

- several maternal and child health and home visitation programs for low-income pregnant women at risk of Zika infection, and families that have children born with birth defects related to Zika infection;
- several health care workforce assistance programs for Puerto Rico and other territories; and
- compensation for persons harmed by the use of tests or vaccines used under emergency authority.[10]

The requested PHSSEF funds could, in consultation with OMB, be transferred to other agencies within HHS or across the federal government. The request stated that this transfer authority is to provide flexibility in response to changing needs. No congressional notification requirement is included.

National Institutes of Health (NIH)

The emergency supplemental request seeks $130 million for the NIH National Institute of Allergy and Infectious Diseases (NIAID) to expand research efforts to characterize the progression and effects of Zika infection and other vector-borne diseases, and to develop vaccines against them.[11] Proposed request language would authorize the NIH Director to transfer funds between NIH accounts. No congressional notification requirement is included.

Food and Drug Administration (FDA)

The emergency supplemental request seeks $10 million for FDA's role in reviewing the safety and effectiveness of medical countermeasures (such as test methods, vaccines, and treatments), and post-market monitoring of such countermeasures if and when they become available.

Medicaid Funding for Territories[12]

The emergency supplemental request would temporarily increase the federal matching rate for Medicaid in the territories. The territories operate Medicaid programs under different rules from those that apply to the 50 states and the District of Columbia. Federal Medicaid funding to the states and the District of Columbia is open-ended, but the territories receive capped annual allotments (i.e., the maximum amount of federal funds available in a year). In addition, the Patient Protection and Affordable Care Act (ACA, P.L. 111-148, as amended) provides the territories with additional federal Medicaid funding to use by September 30, 2019. The territories have a federal medical assistance percentage (FMAP) rate (i.e., federal matching rate) for Medicaid of 55%.[13]

The supplemental request includes a provision that would increase the FMAP rate for the territories to 65% for one year beginning with the first day of the fiscal quarter following enactment. This increased FMAP rate would be available for all Medicaid expenditures, not limited to those provided to treat Zika infection. The federal funding for the increased FMAP rate would not count against the territories' annual federal spending caps or additional ACA funding. The Administration estimates this FMAP rate increase would cause federal Medicaid expenditures to grow by $246 million.[14]

There is some question about how this provision would affect Puerto Rico if it were to exhaust its additional ACA funding prior to FY2019.[15] Depending on the timing of enactment, Puerto Rico might not have access to its full annual Medicaid allotments or additional ACA funding for a portion of the time the provision would be in effect.[16]

Retroactive Reimbursement

The request proposes language that would allow funds provided in the act to be used to reimburse HHS accounts for Zika response expenses incurred prior to enactment.

Transfer Authority

The request proposes language that would allow funds appropriated to HHS in the act to be transferred to other federal accounts, including the Department of Defense, the Environmental Protection Agency, and the Department of Agriculture "to prevent, prepare for, and respond to Zika virus, other vector-borne diseases, or other infectious diseases and related health outcomes, domestically and internationally.... ," following consultation with OMB. No congressional notification requirement is included.

Expanded Definition of "Security Countermeasure"

The request proposes language that would allow the government to support the advanced development and procurement of medical countermeasures against Zika virus through Project BioShield. Currently,

Project BioShield supports only countermeasures against specific chemical, biological, radiological, and nuclear terrorist threats.[17] The proposed expansion is not limited to countermeasures against the Zika virus or vector-borne diseases, but rather is stated broadly as a "countermeasure to diagnose, mitigate, prevent, or treat harm from any infectious disease that may pose a threat to the public health."

International Assistance Programs

The Administration's February 2016 emergency supplemental appropriations request to respond to the Zika outbreak seeks for the Department of State and USAID a total of $376.1 million. This includes funds for control of the disease, prevention, surveillance, evacuating U.S. employees and American citizens, vaccine development, and diagnostic research, among other things. Specifically within the International Assistance section of the request is a request for transfer authority (without a requirement for congressional notification) with certain limitations, reimbursement authority, and hiring of personal services contractors, as well as authorization to use unobligated Ebola balances to combat Zika and other infectious diseases. Also worth noting is that, unlike HHS, funds for international assistance programs have varying periods of availability, as specified below.

Department of State

A total of $41.1 million is requested for the Department of State operations, multilateral assistance within International Organizations and Programs (IO&P), and international security assistance (nuclear research and techniques) as follows:

- *Diplomatic and Consular Programs account (D&CP)—$*14.6 million to remain available until September 30, 2017. Of this amount.
- $8.4 million to support the Office of Medical Services for medical support and possible evacuation under the Chief of Mission authority of at-risk U.S. employees in Zika-affected countries; and
- $6.2 million to support regional coordination efforts and public diplomacy outreach, among other activities.
- *Emergencies in the Diplomatic and Consular Service*—$4 million to remain available until expended to support response efforts, including potential evacuation of U.S. citizens.

- *Repatriation Loans Program*—$1 million to remain available until expended to finance repatriation loans to U.S. citizens who may seek to leave Zika-affected areas or who have been exposed to or have contracted Zika.
- *Nonproliferation, Anti-Terrorism, Demining and Related Programs (NADR)*—$8 million to remain available until September 30, 2017, for additional voluntary U.S. contributions to the International Atomic Energy Agency (IAEA), an autonomous intergovernmental organization related to the United Nations that promotes the safe, secure and peaceful use of nuclear technologies. Funds would support Zika research to develop and deploy nuclear techniques to help accelerate diagnosis, provide related specialized training, and to implement sterile insect projects to suppress mosquito populations.[18]
- *International Organizations and Programs (IO&P)*—$13.5 million to remain available until September 30, 2017, to support Zika response actions taken by UNICEF, the Food and Agriculture Organization, the WHO, and PAHO.

USAID

For the U.S. Agency for International Development, the Administration is requesting $335 million to cover USAID's health programs and implementation expenses:

- *USAID Operating Expenses (OE)*—$10 million to remain available until September 30, 2017, to support Zika response efforts.
- *Global Health Programs (GHP)*—$325 million to remain available until expended to prevent, treat, or respond to the Zika virus and related health concerns, other vector-borne diseases, or other infectious diseases. Multi-year funding commitments are requested to provide incentives for the development of global technologies such as vaccines, diagnostics equipment, and vector control innovations. Anticipated allocations include
  - $100 million to implement vector management and control activities in Zikaaffected countries;
  - $100 million to stimulate private sector research and development of vaccines, diagnostics, and vector control innovations through public-private partnerships;

- $50 million for maternal and child health support in affected and at-risk countries, including training of health care workers; ensuring access to family planning information, services, and methods; providing support for children with microcephaly; and helping pregnant women and their partners have access to personal protection, including condoms and repellant to protect against mosquitoes;
- $25 million for public health communication and behavior change campaigns for affected communities and countries to take actions to protect themselves from Zika and other vector-borne diseases; and
- $50 million to issue Global Health Security Grand Challenges that would call for groundbreaking innovations in diagnostics, vector control, personal protection, community engagement and surveillance, and other tools to address Zika and other infectious diseases, as well as to develop public-private partnerships to accelerate development of innovative tools and practices.

Use of Ebola Balances for Other Infectious Diseases

Within the Department of State and Other International Programs General Provisions (in addition to the General Provisions for the entire request), the supplemental request would authorize the use of unobligated Ebola Funds (Title IX, Div. J, P.L. 113-235), stating: "[Unobligated Ebola funds] shall also be available to respond to the Zika virus and related health outcomes, other vector-borne diseases, or other infectious diseases."

As of January 1, 2016, the Department of State/USAID's unobligated Ebola funds totaled nearly $1.3 billion. Of that total, about $600 million is available until September 30, 2016 (just a few months away), and about $694 million is available until expended.[19] (See the subsequent section, "HHS, State/USAID, and DOD Unobligated Ebola Response Funds.")

Transfer Authority

The Department of State and Other International Programs General Provisions in the supplemental request would allow transfer of State Department-related funds in the request only among State Department-related accounts within the request and transfer of USAID-related funds in the request only among USAID-related accounts. No congressional notification requirement is included.

Notwithstanding Authority

The supplemental's request for notwithstanding authority could allow funds from this or prior acts supporting the U.S. Zika virus response to be expended despite any previously enacted restrictions and conditions on U.S. foreign aid. For example, if enacted, this authority could allow foreign aid to be provided to states that are otherwise restricted by law: those designated as sponsors of terrorism, those with debt arrearage, human rights violators, or states that practice coercive family planning. The Department of State has indicated in the Global Health Program (GHP) section, however, that funds will provide support for "ensuring access to voluntary family planning information, services, and methods."

Direct Hiring Authority/Personal Services Contractors

The General Provisions Title in the request, and also the General Provisions Title for the Department of State and Other International Programs, allows for expedited hiring authority to directly hire staff during critical public health threats, such as Zika, and to enter into contracts with individuals who are experts in Zika-related fields. This measure does not limit direct hiring or personal services contractors only for Zika-related purposes. This authority for direct hiring and personal services contractors could be used in a broader set of public health circumstances than the Zika virus.

HHS, State/USAID, and DOD Unobligated Ebola Response Funds

In December 2014, the Consolidated and Further Continuing Appropriations Act, 2015 (P.L. 113- 235), provided $5.4 billion in emergency supplemental appropriations to HHS, the Departments of State and Defense, and USAID to address the Ebola outbreak that began in West Africa in January 2014.[20] Because these funds were designated as emergency appropriations, they are effectively exempt from spending limits in the Budget Control Act of 2011 (BCA, P.L. 112-25).[21]

On April 6, 2016, the Obama Administration announced its plan to reprogram $510 million of unobligated FY2015 Ebola funding to respond to the Zika virus.[22] HHS Ebola funds may be reprogrammable without additional congressional action (subject to existing restrictions on reprogramming, including notification). This is because the relevant appropriations measures stated the funds are available for Ebola and other infectious diseases. Some have debated whether congressional action is necessary to provide the Department of State and USAID with the authority to reprogram the unobligated Ebola funds, as much of the funding was appropriated with specific language to be used to "prevent, prepare for, or respond to the Ebola disease outbreak."

Table 3 provides, by account, the original appropriated Ebola funds, remaining (unobligated) amounts, the period of funding availability, and purpose of the funds, based on quarterly reports to Congress as required by the law. As of January 1, 2016, unobligated Ebola funds totaled $2.77 billion: $1.46 billion for HHS, $1.29 billion for State/USAID, and $17.3 million for Defense. A portion of the total, $652.9 million—most of which is USAID funding—expires September 30, 2016. Nearly all of the remaining unobligated funds expire September 30, 2019, or are available until expended. *Table 3* does not incorporate the Administration's announced plans to reprogram $510 million of unobligated Ebola funds.

Table 3. FY2015 Emergency Funds Appropriated for Ebola Response and Related Activities, and Unobligated Balances

Amounts are U.S. dollars in millions. Unobligated amounts are as of January 1, 2016

Agency andAccountorActivity	P.L. 113-235[a]	Unobligated Funds	Period of Availability	Purpose(s)
DEPARTMENT OF HEALTH AND HUMAN SERVICES				
CDC:International activities	603.0	369.4	Until Sept.30, 2019	Disease control assistance to affected and neighboring countries.
CDC:Global Health Security	597.0	525.2	Until Sept. 30, 2019	Implementation of Global Health Security Agenda (GHSA) activities.[b]
CDC:Public health emergency preparedness	165.0	11.3	Until Sept. 30, 2019	Domestic preparedness and response activities, including control in health care settings, and procurement for stockpile.
CDC:State and local	255.0	56.0	Until Sept. 30, 2019	Grants to state health departments for surveillance, testing, case management.

Agency andAccountorActivity	P.L. 113-235[a]	Unobligated Funds	Period of Availability	Purpose(s)
CDC:Worker training	0.0	—	—	$10 million for this activity was transferred to NIH. See below.
CDC:Migration/quarantine	119.3	57.4	Until Sept. 30, 2019	Screening and management of entrants from affected countries/regions.
CDC:Other domestic activities	37.0	11.9	Until Sept. 30, 2019	Vaccine trials and other applied public health research.
CDC Subtotal	1,776.3	1,031.2	—	—
ASPR (PHSSEF):Hospital Preparedness Program	208.5	21.2	Until Sept. 30, 2019	Domestic training, PPE, and establishing regional Ebola Treatment Centers (ETCs).
ASPR (PHSSEF):Other prep. and response	352.2	347.4	Until Sept. 30, 2019	Not specified. Could include domestic treatment costs for affected individuals.
ASPR (PHSSEF):BARDA	157.0	3.7	Until Sept. 30, 2019	Research, development, and procurement of vaccines and treatments.
ASPR/PHSSEF Subtotal	717.7	372.3	—	—
NIH, NIAID	238.0	35.2	Until Sept.30, 2016	Research and clinical trials on investigational vaccines and treatments.
NIH, NIAID	10.0	9.0	Until Sept. 30, 2019	Ebola responder safety training, funds transferred from CDC.
FDA	25.0	13.7	Until expended	Development, review, and regulation of vaccines and treatments.
HHS Total	2,767.0	1,461.4	—	—
STATE DEPARTMENT/USAID				
USAID, Operating Expenses	19.0	15.2	Until Sept. 30, 2016	Operating costs to address Ebola outbreak in West Africa, including temporary staffing and technical support.
USAID, Inspector General	5.6	3.4	Until expended	Oversight of Ebola response in West Africa.
USAID, International Disaster Assistance(IDA)	1,436.3	542.4	Until expended	Disaster assistance to address humanitarian needs for West Africa, such as

Table 3. (Continued)

Agency andAccountorActivity	P.L. 113-235[a]	Unobligated Funds	Period of Availability	Purpose(s)
				rapid response, maintaining surveillance, screening, and contact tracing.
USAID, Global Health Programs	312.0	148.0	Until expended	Expanded USAID global health security activities to control infectious diseases and limit spread of Ebola, including surveillance and building lab capacity.
State/USAID, Economic Support Fund	711.7	583.0	Until Sept. 30, 2016	Training and program assistance to prevent economic and government instability during Ebola crisis, including reimbursement for earlier response. Activities include rehabilitation of the water infrastructure, strengthening health information systems, and developing technology to prevent the spread of Ebola.
State, Diplomatic, Consular Programs(D&CP)	36.4	2.2	Until Sept. 30, 2016	Medical support and evacuation capacity,repatriation assistance, and other needs.
State, Repatriation Loans Program	ns	ns	ns	Repatriation loans to U.S. citizens as necessary related to Ebola outbreak. Funding not specified in P.L. 113-235. Explicit transfer authorityis providedforup to $1 millionfrom D&CPinto this account.[c]
State, International Organizations and Programs(IO&P)	ns	ns	ns	Estimated U.S. contributions to UNMEER. Fundingnot specified inP.L. 113-235. Explicit transfer authority is

Agency andAccountorActivity	P.L. 113-235[a]	Unobligated Funds	Period of Availability	Purpose(s)
				providedfor up to $35.3 million from IDA for this account.[c]
State, Contributions to International Organizations (CIO)	ns	ns	ns	Fundingnot specified inP.L. 113-235. Explicit transfer authority is provided forup to $35.3 million from IDA and $50 million from Global Health Programs for this account.[c]
State, Nonproliferation, Anti-terrorism, Demining, and Related Programs	5.3	0.0	Until Sept. 30, 2016	Biosafety and hazardous materials training in affected countries, efforts to mitigate illicit acquisition of Ebola virus and to promote biosecurity practices associated with outbreak response efforts.
State/USAID Total	2,526.3	1,294.2	—	—
DEPARTMENT OF DEFENSE				
Defense/DARPA:Defense -wide research, development, testing, and evaluation (RDT&E)	45.0	0.8	Until Sept.30, 2016	Developing medical countermeasures technologies (e.g., using antibodies from survivors)and shortening vaccine development time.
Defense/CBDP:RDT&E	50.0	16.4	Until Sept. 30, 2017	Research and Development, testingand evaluation.
Defense/CBDP: procurement	17.0	0.1	Until Sept. 30, 2017	Procurement of detection and diagnosticsystems, mortuary supplies, and isolationtransport units.[d]
Defense Total	112.0	17.3	—	—
TOTAL	5,405.3	2,772.9	—	—

Sources: Departmental spend plans and/or quarterly reports for HHS, State/USAID, and Defense, as required by P.L. 113-235, and obtained by CRS; and additional departmental communications.

Note: Amounts may not add due to rounding; "ns" means not specified.

Glossary: ASPR is HHS Assistant Secretary for Preparedness and Response; BARDA is HHS Biomedical Advanced Research and Development Authority; CBDP is Chemical Biological Defense Program; CDC is HHS Centers for Disease Control and Prevention; DARPA is Defense Advanced Research Projects Agency; FDA is HHS Food and Drug Administration; NIAID is NIH National Institute of Allergy and Infectious Diseases; NIH is HHS National Institutes of Health; PHSSEF is HHS Public Health and Social Services Emergency Fund, administered by the HHS Secretary; PPE is personal protective equipment; UNMEER is United Nations Mission for Ebola Emergency Response, and WHO is World Health Organization.

[a] HHS amounts reflect transfers between the funded agencies, as permitted by the law. P.L. 113-235, 128 Stat. 2522, §604, December 16, 2014.

[b] For more information, see CDC, Global Health Security Agenda, http://www.cdc.gov/globalhealth/security/index.htm.

[c] P.L. 113-235, 128 Stat. 2694, §9001.

[d] Explanatory statement accompanying H.R. 83, Congressional Record, vol. 160 (December 11, 2015), p. H9635.

End Notes

[1] For more information, see CRS Report R44368, Zika Virus: Basics About the Disease; CRS In Focus IF10353, Mosquitoes, Zika Virus, and Transmission Ecology; CRS Insight IN10433, Zika Virus: Global Health Considerations; and CRS Report R44385, Zika Virus: CRS Experts.

[2] White House, Office of Management and Budget, Estimate #1 – FY 2016 Emergency Supplemental: Appropriations Request to Respond to the Zika Virus both Domestically and Internationally, February 22, 2016, https://www.whitehouse.gov/omb/budget_amendments. For more information on discretionary spending limits and Ebola funds designated for emergency requirements, see OMB Final Sequestration Report to the President and Congress for Fiscal Year 2015, January 20, 2015, at https://www.whitehouse.gov/sites/default/files/omb/assets/legislative_reports/sequestration/sequestration_final_january_2015_president.pdf.

[3] For more detail, see https://www.opm.gov/blogs/Director/direct-hire-authority/.

[4] As defined in regulation, "The Government is normally required to obtain its employees by direct hire under competitive appointment or other procedures required by the civil service laws. Obtaining personal services by contract, rather than by direct hire, circumvents those laws unless Congress has specifically authorized acquisition of the services by contract." (48 C.F.R. 37.104(a)) Under this authority, federal agencies can quickly contract with individual scientists, physicians, and other experts to aid in response efforts.

[5] Ryan McCrimmon and Jennifer Shutt, "Zika Funding Squabble Engulfs Senate, House and White House," CQ News, April 26, 2016.

[6] OMB, Shaun Donovan, "Taking Every Step Necessary, As Quickly As Possible, to Protect the American People from Zika," OMB blog, April 6, 2016, https://www.whitehouse.gov/omb/blog.

[7] CBO scored the total budgetary effects of these rescissions as reducing budget authority by $622 million, and reducing outlays by $12 million during FY2016. On net, CBO estimates

the budgetary effects of the bill to be $0 in budget authority and an increase of $9 million in outlays for FY2016.

8 Two other proposals for Zika-related supplemental funding were also considered as amendments to these combined appropriations bills (S.Amdt. 3898 and S.Amdt. 3899, both to S.Amdt. 3896). On May 17, the Senate voted not to invoke cloture on these amendments. No further action has occurred as of the date of this report.

9 Erik Wasson, "Obama Administration Updates Zika Spending Request," Bloomberg, April 18, 2016.

10 This compensation program is described in "Covered Countermeasure Process Fund" in CRS Report RS22327, Pandemic Flu and Medical Biodefense Countermeasure Liability Limitation, and HHS, Health Resources and Services Administration, Countermeasures Injury Compensation Program (CICP), http://www.hrsa.gov/cicp/index.html.

11 This refers to infectious diseases that are transmitted by a living organism (a "vector," such as a mosquito), from one host to another.

12 This section contributed by Alison Mitchell, Specialist in Health Care Financing, Domestic Social Policy Division.

13 For more information about the Medicaid program in the territories, see CRS Report R44275, Puerto Rico and Health Care Finance: Frequently Asked Questions, coordinated by Annie L. Mach.

14 The funding for this provision would be provided through a change in mandatory programs (CHIMP), which is a provision in an appropriations act that affects a mandatory spending program.

15 According to HHS, Puerto Rico is projected to exhaust its ACA Medicaid funding by the end of FY2017. HHS, FY2017 Budget in Brief, February, 2017, p. 97, http://www.hhs.gov/sites/default/files/fy2017-budget-in-brief.pdf.

16 If Puerto Rico were to exhaust its ACA Medicaid funding before the end of FY2019, it would have to significantly increase its own Medicaid funding share in order to maintain the current program. This would worsen its current fiscal situation. For more information about this situation, see CRS Report R44095, Puerto Rico's Current Fiscal Challenges, by D. Andrew Austin.

17 For more information, see HHS, "Project BioShield," https://www.medicalcountermeasures.gov/barda/cbrn/projectbioshield-overview/.

18 For more information see Aabha Dixit, IAEA Office of Public Information and Communication, "Nuclear Technique Can Help Control Disease-Transmitting Mosquitoes," February 3, 2016, https://www.iaea.org/newscenter/news/ nuclear-technique-can-help-control-disease-transmitting-mosquitoes.

19 Based on departmental spend plans and/or quarterly reports for HHS, State/USAID, and Defense, as required by P.L. 113-235, and obtained by CRS; and additional departmental communications.

20 This section addresses funds provided in P.L. 113-235 only; it does not track the $88 million appropriated to HHS for Ebola-related activities in the first FY2015 continuing resolution (P.L. 113-164).

21 For more information on discretionary spending limits and Ebola funds designated for emergency requirements, see OMB Final Sequestration Report to the President and Congress for Fiscal Year 2015, January 20, 2015, at https://www.whitehouse.gov/sites/default/files/omb/assets/legislative_reports/sequestration/sequestration_final_january_2015_president.pdf.

[22] OMB, Shaun Donovan, "Taking Every Step Necessary, As Quickly As Possible, to Protect the American People from Zika," OMB blog, April 6, 2016, https://www.whitehouse.gov/omb/blog.

In: The Zika Virus
Editor: Veronica Salazar
ISBN: 978-1-53610-124-9

Chapter 4

EMERGING INFECTIOUS DISEASES: PRELIMINARY OBSERVATIONS ON THE ZIKA VIRUS OUTBREAK*

Timothy M. Persons,
Government Accountability Office

WHY GAO DID THIS STUDY

Emerging infectious diseases constitute a clear and persistent threat to the health and well-being of people and animals around the world. The Zika virus, which at present appears to be primarily transmitted to humans by infected mosquitos, can cause symptoms including fever, rash, and joint pain. A large ongoing outbreak is occurring in Brazil that started in May 2015. As of February 24, 2016, over 100 cases of U.S. travel-associated Zika virus disease cases have been reported.

Due to concerns about its potential impact, you asked GAO to present preliminary observations on the Zika virus. This statement addresses (1) the epidemiology and transmission of the Zika virus disease, including reporting

* This is an edited, reformatted and augmented version of a report, GAO-16-470T, issued by the U.S. Government Accountability Office, March 2, 2016. This report was originally presented as testimony before the House Committee on Energy and Commerce, Subcommittee on Oversight and Investigations for a hearing on "Examining the U.S. Public Health Response to the Zika Virus." Timothy M. Persons is Chief Scientist at the U.S. Government Accountability Office.

on the incidence of disease and what is known about its link to microcephaly; (2) detection and testing methods; (3) methods for mosquito control; and (4) the proposed federal research agenda as it relates to the Zika virus and Zika virus disease.

To report on these questions, GAO reviewed relevant peer-reviewed scientific literature, epidemiological alerts, agency documents, and prior GAO work from 2003-2016 on related topics; consulted experts in the fields of virology, infectious diseases, and vector control, including industry representatives; and interviewed officials of the CDC and NIH.

WHAT GAO RECOMMENDS

GAO is not making recommendations at this time.

WHAT GAO FOUND

While several countries have reported outbreaks of Zika virus disease—which appear to be primarily transmitted to humans by mosquitos—unanswered questions remain regarding the epidemiology and transmission of the disease. Many factors—including a large number of asymptomatic patients and patients with mild symptoms, and a lack of a consistent international case definition of Zika virus disease—complicate understanding of the virus and may hinder responses to the current outbreak. For example, an estimated 80 percent of individuals infected with the Zika virus may not manifest clinical symptoms. As a result, incidence of the infection may be underestimated. Questions also remain regarding the strength of the association between Zika virus infection and two other conditions: microcephaly and Guillain-Barré syndrome.

A lack of validated diagnostic tests, consistent international case definitions, and trend information may also contribute to difficulty in estimating the prevalence of the virus. The United States uses two diagnostic tests for Zika, and according to the U.S. Centers for Disease Control and Prevention (CDC), while there are no commercially-available diagnostic tests for Zika, an antibody-based test for Zika virus was recently authorized for Emergency Use by the U.S. Food and Drug Administration. Diagnosing Zika virus infection is also complicated because it is difficult to differentiate it from other similar diseases, such as dengue or yellow fever. For example, a person previously infected with dengue could be falsely identified as also having been exposed to the Zika virus (and vice-versa). Moreover, the World Health Organization has acknowledged the need for a consistent case definition—that is, a set of uniform criteria to define the disease for public health surveillance and to determine who is included in the count and who is excluded. Additionally, a lack of pattern and trend data has made surveillance challenging.

Because Zika virus disease cannot yet be prevented by drugs or vaccines, vector (mosquito) control remains a critical factor in preventing and mitigating the occurrence of this disease. There are three methods for mosquito control: (1) standing water treatment, (2) insecticides, and (3) emerging technologies. Mosquito control has been achieved in some locations by methods such as reducing or chemically-treating water sources where mosquitoes breed or mature, or by insecticide dispersal. Emerging technologies, including biological control methods—such as infecting mosquitoes with bacteria—genetically-modified mosquitoes, and auto-dissemination traps, show some promise but are still in development and testing phases.

The National Institutes of Health (NIH) and the CDC have identified several high priority areas of research. Research priorities include basic research to understand viral replication, pathogenesis, and transmission, as well as the biology of the mosquito vectors; potential interactions with co-infections such as dengue and yellow fever viruses; linkages between Zika and the birth defect microcephaly; improving diagnostic tests; vaccine development; and novel vector control methods. These efforts are ambitious, and agencies may face challenges in implementing this agenda.

* * *

Chairman Murphy, Ranking Member DeGette, and Members of the Subcommittee:

I am pleased to be here today to discuss our preliminary observations on Zika virus disease. Zika virus—which at present appears to be primarily transmitted to humans by infected mosquitoes—can cause symptoms including fever, rash, conjunctivitis (red eyes), and joint and muscle pain. Although a possible link between Zika virus disease and microcephaly in newborn babies, as well as a link with Guillain-Barré syndrome in infected adults has been hypothesized, to date, the link has not been firmly established.[1] The

Government Accountability Office (GAO) has issued several reports dealing with preparedness for, detection of, and response to emerging infectious diseases domestically and internationally, as well as pandemics. (See appendix 1 for a listing of these reports).

Emerging infectious diseases represent an ongoing threat to the health and livelihoods of people and animals worldwide.[2] Many advances in medical research and treatments have been made during the last century, but infectious diseases are nevertheless a leading cause of death worldwide. In addition to causing nearly one in five human deaths worldwide, infectious diseases impose a heavy societal and economic burden on individuals, families, communities, and countries.[3] Infectious diseases are a continuous threat for reasons that include: (1) emergence—at times rapid—of new infectious diseases; (2) reemergence of previously-known infectious diseases; and (3) persistence of intractable infectious diseases. According to the National Institutes of Health (NIH), infectious diseases continue to emerge and evolve. Changes in human demographics, behavior and land use—among other factors—bring people into closer and more frequent contact with pathogens and are contributing to infectious disease emergence. This may involve exposure to animal or insect carriers of disease. There is also an increase in the opportunities for pathogens to jump between animal and human reservoirs. In addition to Zika, other examples of emerging infectious diseases include: Ebola virus disease, severe acute respiratory syndrome (SARS), influenza, dengue, and chikungunya, among others.

Due to concerns about the potential impact of Zika virus disease, you asked us to present our preliminary observations on several issues associated with Zika virus disease.[4] This statement addresses issues, if any, associated with (1) the epidemiology and transmission of the Zika virus disease, including reporting on the incidence of disease and what is known about its link to microcephaly; (2) detection and testing methods; (3) methods for mosquito control; and (4) the proposed federal research agenda as it relates to the Zika virus and Zika virus disease.

To report on these questions, we reviewed relevant peer-reviewed scientific literature, epidemiological alerts, agency documents, and prior GAO work from 2003-2016 on related topics such as disease surveillance and emerging infectious diseases; consulted experts in the fields of virology, infectious diseases, and vector (mosquito) control, including industry representatives; and interviewed officials of the Centers for Disease Control and Prevention (CDC) and the National Institutes of Health (NIH).

The work upon which this testimony is based was conducted from February through March 2016 in accordance with generally accepted government auditing standards. Those standards require that we plan and perform the audit to obtain sufficient, appropriate evidence to provide a reasonable basis for our findings and conclusions based on our audit objectives. We believe that the evidence obtained provides a reasonable basis for our findings and conclusions based on our audit objectives.

BACKGROUND

Zika virus is a member of the flavivirus family related to dengue virus, yellow fever virus, and West Nile virus, and its primary mode of transmission is via mosquito, most notably by the *Aedes aegypti* mosquito.[5] *Aedes albopictus* mosquitoes are a potential vector for the Zika virus. The disease was first identified in the Zika Forest in Africa in the 1940s, from which it subsequently moved eastwards over the following decades, though the Pacific Islands until it reached Brazil, where the population had no indigenous immunity against the virus. According to CDC officials with whom we spoke, the United States also has no indigenous immunity to the Zika virus.

A major factor contributing to the declaration of Zika virus disease as a Public Health Emergency of International Concern by the World Health Organization (WHO) is a possible link between Zika virus and microcephaly as well as Guillain-Barré syndrome. Microcephaly—an abnormally small head due to failure of brain growth—is a concern because children with microcephaly can experience impaired cognitive development, delayed motor function and speech, seizures, and reduced life expectancy.

To better understand the spread of this disease worldwide, epidemiological studies are required. Epidemiology is concerned with the patterns of disease occurrence in human populations and of the factors that influence these patterns. The goals of epidemiologic study, and more specifically outbreak investigations, are to determine the extent and distribution of the disease in the population, the causes and factors associated with the disease and its modes of transmission, the natural history of disease, and the basis for developing preventive strategies or interventions[6] (see figure 1).

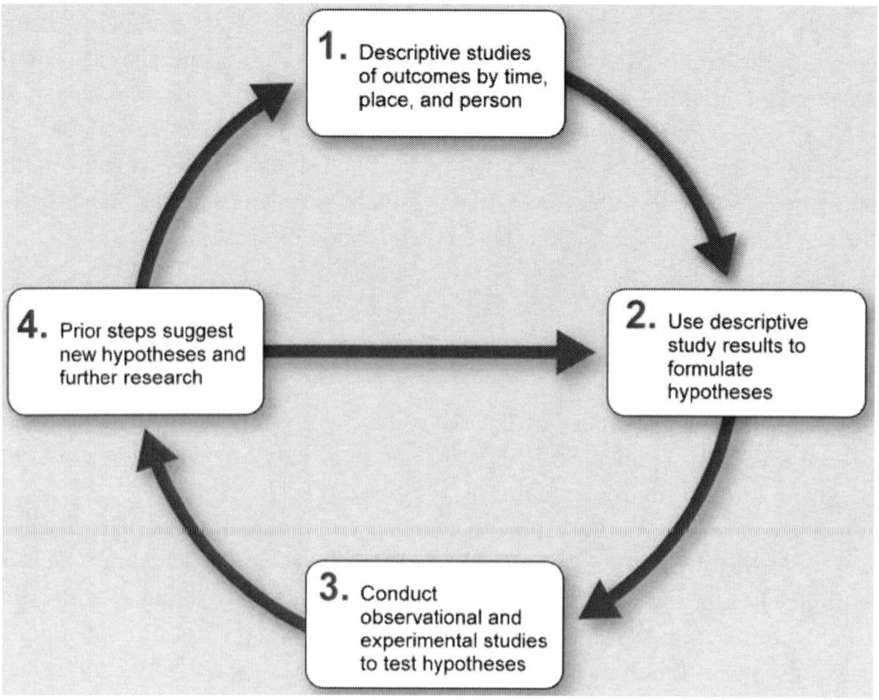

Source: GAO, based on concepts regarding the steps in conducting epidemiological studies drawn from Lilienfeld, 2015. GAO-16-470T.

Figure 1. Epidemiological Study Cycles.

U.S. Role in Global Disease Surveillance

The United States has played a significant role in improving global disease surveillance and response capacity. In the mid-1990s, recognizing the threat posed by previously unknown infectious diseases, the United States and other countries initiated a broader effort to ensure that countries can detect disease outbreaks that may constitute a Public Health Emergency of International Concern. The United States has participated in the WHO's efforts to develop and implement the International Health Regulations, currently an agreement among 196 countries, to develop and maintain global capabilities to detect and respond to disease and public health threats. The CDC has helped to define and establish the International Health Regulations and has been designated by the WHO as a key partner in helping to implement the critical capacities for detecting and responding to emerging infectious disease outbreaks.

The recent Ebola outbreak in West Africa has highlighted the importance of further improving the U.S. government's global disease surveillance efforts. To help ensure that such threats are addressed early and at their source, the "National Health Security Strategy and Implementation Plan 2015-2018" released by the U.S. Department of Health and Human Services prioritizes efforts to strengthen national capacities and capabilities globally to detect disease in a timely manner, prevent the global spread of public health threats and diseases, and respond to public health emergencies.[7] For example, CDC' s Global Disease Detection and Field Epidemiology Training programs aim to strengthen laboratory systems for the rapid detection and control of emerging infectious diseases and train epidemiologists to effectively detect, investigate, and respond to health threats.

The Epidemiology and Transmission of the Zika Virus Are Not Fully Understood and Accurate Information on the Incidence Is Lacking

While several countries have reported outbreaks of Zika virus disease, unanswered questions remain regarding the epidemiology and the transmission of the disease. Many factors, including a large number of asymptomatic patients, mild symptoms, a lack of a consistent international case definition of Zika virus disease, as well as of microcephaly, and a lack of validated diagnostic tests complicate our understanding of the virus and may hinder our response to the current outbreak. Questions also remain regarding the strength of the association between Zika virus infection and microcephaly or Guillain-Barré syndrome.

Several Countries Have Reported Outbreak of Zika Virus Disease

Since the 1960s, the Zika virus has been known to occur within a narrow equatorial belt from Africa to Asia. In 2007, the virus was detected in Yap Island, the first report that the virus spread outside of Africa and Indonesia to Pacific Islands.[8] In 2014, the virus spread east across the Pacific Ocean to

French Polynesia, then to Easter Island. According to the WHO, the virus has continued to spread to the Americas, with the outbreak in Brazil that began in May 2015 and is ongoing. Zika has spread to Mexico, Central America, the Caribbean, and South America, where the outbreak has reached epidemic levels (see figure 2). Recent outbreaks have also been reported in Puerto Rico, as well as the Cape Verde Islands.

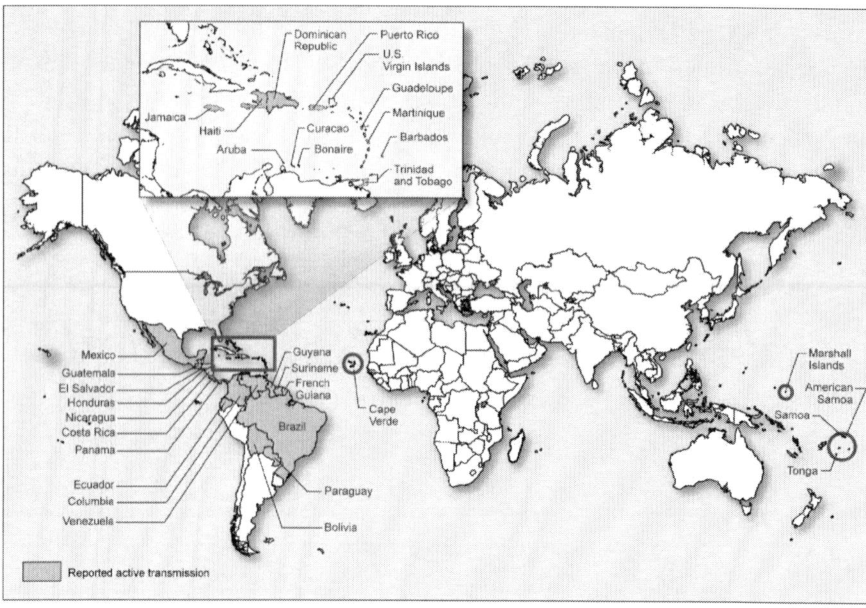

Sources: U.S. Centers for Disease Control and Prevention (CDC), 2016; Map Resources (map). | GAO-16-470T.

Note: This map depicts local mosquito-borne transmission – infections originating in the place where they are found. This is different from travel-associated infections which originate in one place, and are found in another.

Figure 2. All Countries and Territories with Active Zika Virus Transmission.

A Number of U.S. Travelers Have Been Infected with the Zika Virus

According to CDC documentation, Zika virus disease is now a nationally notifiable disease.[9] As of February 24, 2016, 107 cases of continental U.S. travel-associated Zika virus disease have been reported, according to CDC. Although CDC documentation states that Zika virus has not yet seen local mosquito-borne spread in the continental United States, some states have mosquito species potentially capable of transmitting the virus.

Incidences of mosquito-borne transmission have been reported in the Commonwealth of Puerto Rico, the U.S. Virgin Islands, and American Samoa. The first locally-acquired case of Zika virus disease in Puerto Rico was reported in December of 2015. Through late January of 2016, about 30 additional laboratory-confirmed cases were identified in Puerto Rico, including one pregnant woman. In January of 2016, the CDC issued travel guidance for travel to affected countries, including the use of enhanced precautions for all travelers, as well as the recommendation that pregnant women postpone travel to affected areas.

Questions Remain about the Transmission of the Zika Virus

According to CDC documentation, there are a few known routes of transmission of the Zika virus to and among people. These include mosquitoes, mother to child, sexual contact and blood transfusions. The Zika virus is transmitted to people primarily through the bite of infected *Aedes* species mosquitoes (primarily *Aedes aegypti* and possibly *Aedes albopictus*). These are the same mosquitoes that spread dengue and chikungunya viruses. These mosquitoes typically lay eggs in and near standing water in containers like buckets, bowls, animal dishes, flower pots, and vases. They prefer to bite people, and live both indoors and outdoors. Mosquitoes that spread dengue, chikungunya, and Zika are aggressive daytime biters, but also bite at night. Mosquitoes can become infected when they feed on a person already infected with the virus.

According to the CDC, the Zika virus is rarely transmitted from an infected mother to child, and there have been no reports of infants contracting the Zika virus through breastfeeding. It is possible but rare that an infected mother would pass the virus to a newborn at delivery. However, an infected mother can pass the Zika virus to her fetus during pregnancy.

According to the CDC, it is possible for the Zika virus to be spread by a man to his sexual partners. A few recent cases of Zika virus transmission were reported through sexual contact. In December 2013, during a Zika virus outbreak in French Polynesia, Zika was isolated from the semen of a patient.[10] In one known case of likely sexual transmission, the virus was spread before symptoms developed. The virus appears to be present in semen longer than in blood. Sexual transmission of the disease— acquired outside of the United

States—has been reported in the United States.[11] As of February 23, 2016, the CDC and state public health departments are investigating 14 additional reports of possible sexual transmission of the virus, including several involving pregnant women.

Zika virus can also be transmitted through blood transfusion, according to U.S. Food and Drug Administration (FDA) documents. While there have been no reports to date of Zika virus entering the U.S. blood supply, the risk of blood transmission is considered high based on the most current scientific research of how Zika virus and similar viruses (i.e., flaviviruses) are spread, as well as recent reports of transfusion-associated infection outside of the United States, according to the FDA.[12] CDC reports that there have been reports of possible blood transfusion transmission cases in Brazil. During the French Polynesian outbreak in 2013, 2.8 percent of blood donors tested positive for Zika. The maximum time the virus remains in the bloodstream is unknown, but scientists estimate that it is less than 28 days. On February 16, 2016, as a safety measure against the emerging Zika virus outbreak, the FDA issued new guidance recommending that blood donors be deferred for four weeks if they have been to areas with active Zika virus transmission, potentially have been exposed to the virus, or have had a confirmed Zika virus infection.

While scientific studies have identified Zika viral components in saliva and urine, they did not report disease transmission from those bodily fluids.

Questions Remain Regarding the Link between Zika Virus Infection and Neurological and Auto-Immune Disorders

It is not currently known if infection with the Zika virus causes, facilitates, or is otherwise associated with the development of certain neurologic and auto-immune conditions. Although strongly suspected, a report suggests the causal relation between in-utero exposure to Zika and microcephaly is yet to be established.[13] Scientific literature has identified several possible linkages, including the presence of Zika virus in fetal brain tissue, as well as evidence of the virus crossing the placental barrier, suggesting a causal effect is plausible, but not yet proven. For example, a fetal autopsy identified an abnormally small brain, as well physical markers of developmental delays, along with the Zika virus in the brain. The mother in the case reported an illness with a fever and rash at the end of the first trimester of pregnancy while she was living in Brazil.[14]

A retrospective analysis was reported to the WHO in 2015-2016 of a previous outbreak of Zika virus disease in 2013-2014 in French Polynesia, also established elevated numbers of neurological disorders for that outbreak.

The potential association of Guillain-Barré syndrome and Zika virus disease was suspected prior to the recent Brazilian Zika disease outbreak. According to the European Center for Disease Control, the 2013 to 2014 French Polynesian outbreak of Zika virus disease was reportedly the largest documented outbreak at that time. According to the European Center for Disease Control, over 8,000 suspected cases of Zika Virus infection had been reported by February 2014.[15] Notably, there were nearly 40 cases of Guillain-Barré syndrome reported, with all cases following disease episodes compatible with Zika virus infection. Historically, there had been 10 or fewer cases of Guillain-Barré syndrome in Polynesia annually. The European Center for Disease Control stated that further investigations could be conducted to establish the relationship between neurological and auto-immune complications and Zika virus infection

Mild Symptoms and Asymptomatic Cases May Lead to Underestimation

Researchers have reported that an estimated 80 percent of the individuals infected with the Zika virus are asymptomatic, that is, they have the virus but do not manifest clinical symptoms. Since diagnosis of suspected Zika virus disease is often based on clinical symptoms, and in light of the fact that clinical symptoms are usually non-existent or mild, experts told us that many individuals who are infected with the Zika virus may not seek medical care, and thus are not counted as a case, resulting in significant underestimation of the true incidence of infection.

Lack of Consistent International Case Definitions May Lead to Inaccurate Count

An accurate count of the number of Zika virus disease cases requires a consistent case definition, or set of uniform criteria to define the disease for public health surveillance and to determine who is included in the count and who is excluded. However, establishing a definition is problematic for Zika virus disease. If Zika cases are diagnosed based on serology data, then the incidence count may include people who have been infected with the virus, but do not show clinical symptoms.[16] On the other hand, if cases are defined by clinical symptoms only, with no serology testing, then the incidence could be higher or lower than that count obtained from serology testing only, because people who present with clinical symptoms may or may not actually test positive for Zika virus.

According to the WHO Zika Response Strategy, there is currently a need to establish a uniform case definition for Zika virus disease, as well as historical rates, or baselines, for associated conditions. The Council of State and Territorial Epidemiologists currently has a case definition for Zika virus disease—under arboviral diseases—with two tiers: a "probable case" definition based on clinical signs and symptoms as well as the presence of certain anti-Zika antibodies, and a "confirmed case" definition for laboratory-confirmed cases based on laboratory analysis.[17] However, because other countries may be using different testing protocols, it is unclear whether their results would be consistent with the CDC case definitions, complicating epidemiological analysis. Engaging international cooperation to establish uniform case definitions and baselines for diagnosing Zika virus disease and microcephaly can facilitate discovery of modes of transmission and causal links between Zika virus disease and microcephaly or Guillain-Barré syndrome.

Lack of a Case Definition of Microcephaly May Complicate Linkage with Zika Virus Disease

When the WHO declared a Public Health Emergency of International Concern on February 1, 2016, it acknowledged that there was no international standard surveillance case definition for microcephaly.[18] Problems with changing case definitions, lack of sufficient information on underlying causes and brain pathology, and lack of baseline data make it difficult to accurately determine the level of increase of microcephaly in Brazil, and how much is due to the Zika virus.[19]

Some researchers offered several possible explanations for the observed increase in microcephaly cases, other than actual increase of cases as a result of Zika virus infection. First, because of the recent attention, newborn babies with visible cranial deformities are likely to be fast-tracked for in-depth examination. This temporal increase in suspected cases of microcephaly could also be distorted given both raised awareness with more children than usual being measured and reported, and the changing definition of microcephaly over time. Although there is evidence of an increased number of cases of microcephaly in Brazil, these authors demonstrated that the number of suspected cases relied on a screening test that had very low specificity and therefore overestimated the actual number of cases.

Lack of Validated Diagnostics Tests May Mask the Incidence of Zika Virus Disease

According to the CDC, there are currently two Zika diagnostic tests available in the United States: reverse transcription polymerase chain reaction (RT-PCR) and Immunoglobulin M (IgM) followed by the Plaque Reduction Neutralization Test (PRNT) test.[20] The current RT-PCR test can detect infection only during the period of illness when the virus is present. According to an NIH official, the PRNT diagnostic test is the most specific for antibody detection,[21] but is cumbersome and not suitable for screening a large number of individuals.

When detecting antibodies, diagnosing cases of Zika virus disease and differentiating it from other diseases caused by other flaviviruses, such as dengue or yellow fever, is difficult if someone has been infected by another flavivirus. Some tests for Zika virus antibodies suffer from cross-reactivity with antibodies to similar viruses, such as from dengue virus disease, meaning that tests using these antibodies for detection are not specific for the Zika virus. For example, a person previously infected with another flavivirus such as dengue could be falsely identified as also having been exposed to the Zika virus (and vice-versa). In addition, new outbreaks like Zika may not have known patterns or trends, making effective surveillance challenging. For example, after the onset of illness, Zika virus remains in the blood for about 5-7 days, according to the CDC. After this period, called viremia, diagnosis of Zika virus disease at this time relies on detection of antibodies against the Zika virus. Since antibodies in the blood may persist longer than the virus, a positive result for antibodies against the Zika virus indicates only that the patient was previously exposed to the Zika virus. Thus, the window for detecting the actual virus is small. It is not clear whether an antibody test could determine how long ago the patient was exposed.

According to the CDC, while there are no commercially-available diagnostic tests for Zika, an antibody-based test for the Zika virus (Zika MAC-ELISA) was recently authorized for Emergency Use by the FDA. One of the main limitations of this test, among others, is its inability to differentiate between infection with Zika and other closely related flaviviruses such as dengue. This test in its current form may confuse the practitioners because of its lack of specificity. Since closely related flaviviruses such as dengue may also be present in Zika outbreak countries, the utilization of this assay could wrongly identify non-Zikavirus associated infections, thus putting extra burden on the laboratory and health care systems, and distort the epidemiological analyses.

Adding to the limitations of these diagnostic systems are limited numbers of facilities able to perform definitive confirmatory testing, particularly in the developing world. The WHO is undertaking an analysis of diagnostics under development, developing target product profiles, facilitating the preparation and characterization of reference reagents, and setting up an Emergency Use Assessment and Listing mechanism for priority Zika diagnostics.

Various Methods of Mosquito Control Exist and Novel Approaches Are Underway

Because Zika virus disease cannot yet be prevented by drugs or vaccines, vector (mosquito) control remains a critical factor in mitigating risks associated with this disease. The *Aedes aegypti* and *Aedes albopictus* mosquitoes are present around the world, as well as in the United States. Figure 3 shows a predicted distribution and intensity of the Aedes aegypti and Aedes albopictus mosquitoes, respectively, and indicates that the southeastern United States, particularly the Gulf Coast states, could be at risk of exposure in the near future. Figure 4 shows the approximate distribution of *Aedes aegypti* and *Aedes albopictus* mosquitoes in the United States in more detail.

There are both large scale and personal methods for mosquito control. We provide a brief preliminary overview of some population-scale control methods identified in the literature, agency documents, and interviews with industry officials and academicians, which include three potentially-overlapping categories: (1) standing water treatment, (2) insecticides, and (3) emerging technologies. For some of these emerging techniques, their effectiveness remains to be demonstrated, but they have the potential to be additional tools in mosquito control. Personal methods, such as use of repellents, nets or long-sleeved clothing and staying indoors in air-conditioned locations are out the scope of this report.

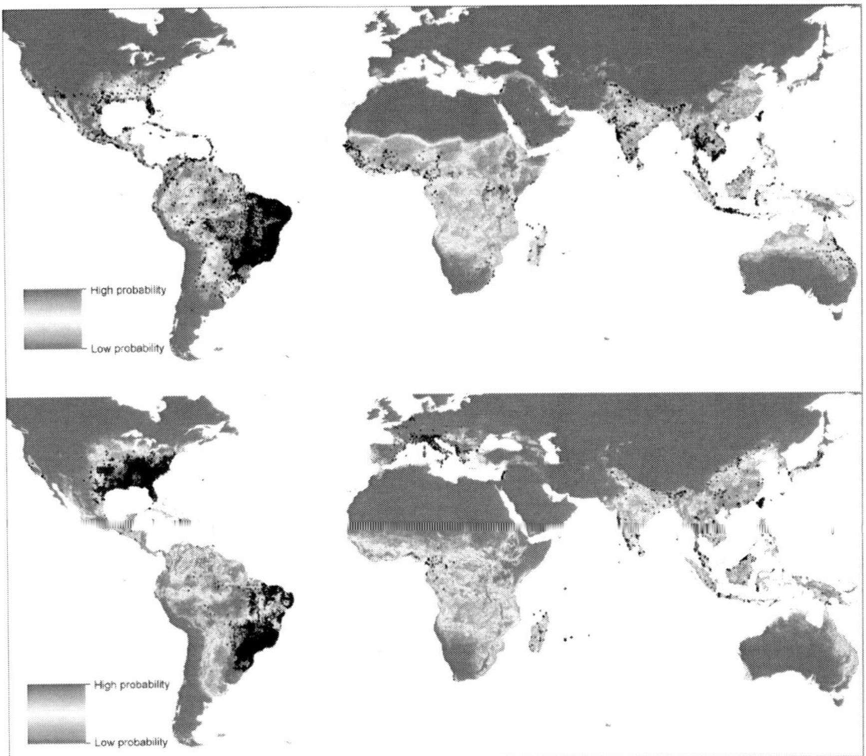

Source: Moritz UG Kraemer, University of Oxford, http://dx.doi.org/10.7554/eLife.08347; MapResources (map). | GAO-16-470T.

Note: Black dots represent a measured occurrence of each mosquito species and colored shading represent an estimated or projected distribution of each mosquito on a spectrum from low probability (blue) to high probability (red).

Figure 3. The actual and predicted distribution of two species of mosquito that can transmit Zika virus: *Aedes aegypti* (top) and *Aedes albopictus* (bottom).

Standing Water Treatment

Standing water treatment for mosquito control can be achieved by the physical reduction of bodies of water or by treating the water with chemicals that kill mosquito larvae or interfere with their development. According to CDC documentation, these treatments include use of certain bacteria or insecticides that mimic mosquito hormones, which prevents mosquitoes from maturing or kills them as larvae. Another method involves coating water with a thin film of oil to suffocate immature larvae.

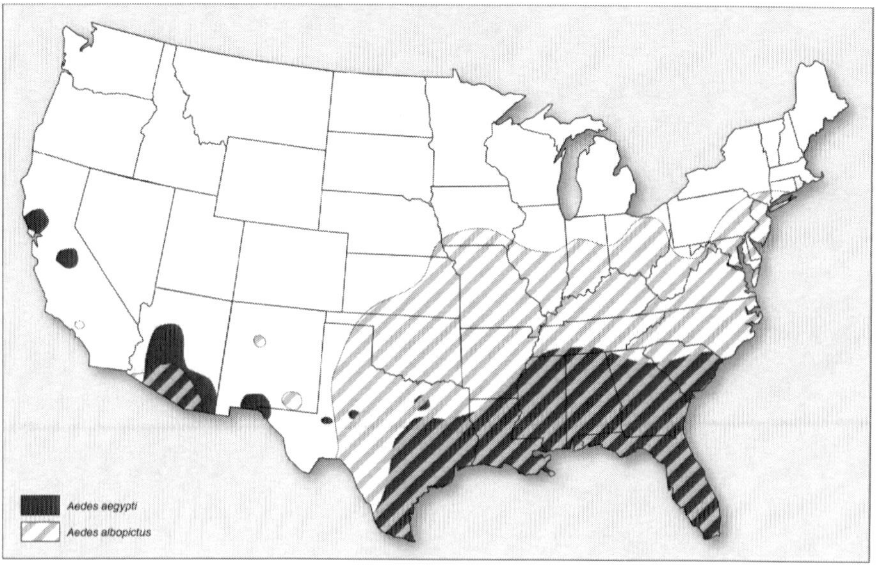

Sources: U.S. Centers for Disease Control and Prevention (CDC); Map Resources (map). | GAO-16-470T.

Note: Maps were developed by CDC using currently available information (as of January 2016). Mosquito populations may be detected in areas not shaded on this map, and may not be consistently found in all shaded areas.

Figure 4. Approximate distribution of *Aedes aegypti* and *Aedes albopictus* mosquitoes in the United States.

Insecticide Dispersal

According to CDC documentation, insecticide dispersal relies on various techniques such as space spraying by trucks or aircraft, or residual spraying, which entails coating surfaces with insecticide. Spraying is one method currently used in the United States. CDC documentation and scientific literature have established that in the long-term, the effectiveness of spraying may be diluted due to insecticide resistance, concerns over environmental exposure, and questionable efficacy of externally-delivered wide-area fogging or spraying. Additionally, the WHO notes that reactive space spraying during emergencies has a low impact unless integrated with other control strategies.

Researchers are developing new chemicals that are more targeted towards mosquitoes, and are attempting to alleviate human toxicity issues. [22]

The use of insecticides as control methods effectively reduced mosquito-borne diseases, including malaria and yellow fever, in most of the world in the 1940s-1960s. The WHO determined that maintaining vector control after a disease subsides is complicated by dwindling resources. Indeed, by the 1980s-1990s, many dangerous vector-borne diseases re-emerged or spread to new regions.

Additionally, the spread of some diseases, such as dengue virus disease, can be attributed to a combination of mosquito, viral, and human factors. To address the resulting complexities of such disease transmissions, the WHO uses "Integrated Vector Management," which leverages multiple control methods based on surveillance and evaluation of involved insects and disease epidemiology.

Emerging Technologies

Emerging technologies include (1) use of biological control methods, (2) genetically-modified mosquitoes, and (3) auto-dissemination traps. Based on scientific literature, these technologies show some promise in studies to overcome issues associated with use of insecticides, such as insecticide resistance.[23] Many of these methods have been tested in smaller-scale controlled field trials internationally. We have not done an independent, comprehensive evaluation of these technologies, due to time limitations.

Biological Control Methods

Biological control methods include introducing natural predators of mosquitoes or their eggs and using bacteria to prevent disease transmission to humans. For example, certain small crustaceans and certain fish eat mosquito larvae. The suitability of this approach for the Zika virus is uncertain because the primary mosquito species identified for Zika transmission are the *Aedes aegypti* and *Aedes albopictus*, which can breed in very small volumes of water, such as those found in tin cans or in plates under potted plants.

Scientific literature has identified another approach—using bacteria to reduce disease transmission from mosquitoes to humans. This bacteria, called *Wolbachia*, can be transferred from mosquitoes to their eggs, thereby propagating this effect to future generations. This tactic has been demonstrated in laboratory environments and is undergoing field trials internationally, particularly in areas affected by the dengue virus.

Genetically Modified Mosquitoes

Control of the disease-transmitting mosquito population using genetically modified mosquitoes can be potentially achieved in different ways.[24] Some genetically-modified mosquitoes are engineered with a "lethal" gene that constantly makes a protein that kills the mosquito larvae. According to one company creating these mosquitoes, the use of genetically modified mosquitoes allows for population control by introducing male genetically-modified mosquitoes that transfer this lethal gene to the female mosquito' s eggs. As a result, mosquito larvae with the inherited lethal gene die. The company claims it has achieved a 90 percent reduction in mosquito populations using its method of releasing the modified male mosquitoes 1-3 times weekly over a period of months. The modified mosquitoes do not persist in the environment, as released mosquitoes generally die out on their own. Given the estimated 200- meter range of a male mosquito during its lifetime, scalability may be challenging. Another genetically-modified mosquito method incorporates viral resistance into mosquitoes, with the goal of replacing existing populations of mosquitoes with one less capable of disease transmission, rather than reducing the number of mosquitoes. Scientific literature indicates there may be public opposition to release of genetically-modified mosquitoes for either procedure due to uncertainties about their effect on people and the environment, or other unknown consequences.

Auto-Dissemination Traps

Mosquito control by auto-dissemination traps functions similarly to insecticides, but relies on containers coated with similar chemicals. After a mosquito lands on these containers, they are contaminated with the chemicals and subsequently transfer them to the location where they lay eggs, which may result in larval death. In the case of the *Aedes aegypti* mosquito, WHO documentation indicates these locations tend to be small containers, so auto-dissemination is particularly suited to these mosquitoes. Auto-dissemination traps have shown 42-98 percent decreases in *Aedes aegypti* mosquito population in field trials.

FEDERAL PRIORITY RESEARCH AGENDA IS AMBITIOUS AND IMPLEMENTATION OF SOME RESEARCH MAY BE CHALLENGING

NIH and CDC have identified areas of high priority for research related to the Zika virus. In addition, the Administration's $1.9 billion emergency funding request to combat the spread of the Zika virus is intended to provide resources to both NIH and CDC to this end. If approved, the NIH would receive $130 million to support vaccine development and other work related to Zika and other mosquito-borne diseases such as chikungunya. The CDC would receive $828 million, including $225 million for grants and technical assistance, in part to expand mosquito or vector control. The FDA would receive $10 million, in part to support the approval of new diagnostic tests and evaluation of treatment efficacy.

NIH has identified areas of high priority including:

- Basic research to understand viral replication, pathogenesis, and transmission, as well as the biology of the mosquito vectors; potential interactions with co-infections such as dengue and yellow fever viruses; animal models of Zika virus infection; and novel vector control methods; and
- Pursuing Zika virus research to develop sensitive, specific, and rapid clinical diagnostic tests; drug treatments for Zika virus as well as broad spectrum therapeutics that treat multiple flaviviral infection; and effective vaccines and vaccination strategies.

On February 5th, 2016, several NIH institutes issued a notice to researchers indicating NIH' s interest in supporting research to understand transmission of the Zika virus, optimal screening and management in pregnancy, and the mechanisms by which the Zika virus affects the developing nervous system, including potential links to microcephaly. This notice was followed by a Funding Opportunity Announcement issued on February 19th, 2016, to create an expedited mechanism for funding exploratory and developmental research projects on these topics.[25]

The CDC has identified priority areas of research focus on Zika including:

- Determining the link between Zika virus infections and the birth defect microcephaly and measuring changes in incidence rates of the birth defect over time;
- Improving diagnostics for the Zika virus, including advanced methods to refine tests and support advanced developments for vector control; and
- Enhancing international capacity for virus surveillance, expanding laboratory testing, and health care provider testing in countries at highest risk of Zika virus outbreaks.

These research activities are intended to supplement other activities for response, readiness and surveillance.

The priority research areas identified by NIH and CDC are ambitious and agencies may face some challenges in implementing this agenda, including:

- Given that there are few known cases in the United States, NIH and CDC may have to rely on the cooperation of other countries with sufficient number of cases in order to carry on the proposed research. However, data from other countries may be different due to different definitions of Zika virus disease and microcephaly.
- Demonstrating the link between the Zika virus and microcephaly may depend not only on the presence of the virus, but also on environmental and nutritional factors.[26] In addition, shifting case definitions and a lack of baseline data makes it difficult to determine the increase, if any, in microcephaly and how much can be attributed to the Zika virus.
- The presence of a high percentage of asymptomatic cases makes it difficult to conduct epidemiological studies, both in identifying exposed and unexposed individuals for case control studies.[27]
- Prior infection or co-infection with another virus such as dengue may complicate any analyses.

NIH officials told us that prior work on similar viruses has allowed them to make rapid progress on both a DNA-based vaccine (based on prior work on West Nile Virus), and a live attenuated virus vaccine modeled after the dengue virus vaccine that is currently in phase 3 clinical trials in Brazil. NIH officials told us that prior "platform" work on similar viruses has expedited response to the Zika outbreak, and in the development of diagnostic tests and vaccines.

With regard to the dengue vaccine, NIH provided us with the following timeline:

Pre-clinical Development	Began 1999
Phase I Clinical Trial	2000 - 2012
Phase II Clinical Trial	2012 - present
Phase III Clinical Trial	2015 - Present

It has taken more than 16 years, and trials are not yet complete. NIH officials told us that given their past experience with the development of vaccine for dengue fever, a vaccine for Zika could be ready for use in an emergency situation in three to four years, in the best case scenario. However, when we asked NIH about this estimate, NIH stated that the National Institute of Allergy and Infectious Diseases plans to begin a Phase I clinical trial within this calendar year. If a candidate vaccine shows promise in Phase I testing, additional clinical testing could begin by 2017 in countries where the disease is found, if the outbreak is still ongoing. The progress of these additional tests, and whether they can contribute to successful licensure, depends on a number of factors, including scientific and technical progress, as well as the size of any ongoing Zika outbreaks during clinical testing. For this reason, it is difficult to provide an exact estimate for the time it will take to develop a Zika vaccine from preclinical studies through clinical testing and licensure.

Zika virus disease poses new challenges to vaccine development and testing. This disease has specific and important implications for pregnant women. There are substantial knowledge gaps in current understanding of Zika, irrespective of the affected population. Since Zika virus disease is associated with, and may cause, adverse fetal outcomes, pregnant women are at particular risk and may benefit from measures such as vaccines. There are several current scientific and structural barriers to developing and testing vaccines for pregnant women.[28] Overcoming these barriers may extend timeframes for vaccine testing and approval. The information we have from NIH and our prior work suggests that development of a Zika virus vaccine may take longer than anticipated by NIH.

SUMMARY

Though technological and scientific advances have improved outbreak detection and response in recent decades, GAO's past work has demonstrated that significant challenges remain in addressing emerging infectious diseases, including a lack of basic public health surveillance and response capabilities in many regions, which can delay and compromise effective detection, diagnosis, and treatment (including vaccine development and therapeutics).

Most of the recent infectious diseases have origins in other parts of the world and have thrived elsewhere long before entering the United States. GAO's past work on the swine enteric coronavirus disease virus in pigs showed the lack of basic preparedness for an emerging animal disease. This, and observations on the Zika virus, point to a persistent and urgent need for a proactive, agile, integrated, and coordinated set of programs concerning research and development, including epidemiological studies, mosquito control, testing capabilities, modeling and simulation, and vaccine development, manufacturing, deployment, biosurveillance, risk management, laboratory capacity expansion, and pandemic mitigation.

Chairman Murphy, Ranking Member DeGette, and Members of the Subcommittee, this concludes my prepared statement. I would be happy to respond to any questions you may have.

Appendix I: Related GAO Work

Biosurveillance: Ongoing Challenges and Future Considerations for DHS Biosurveillance Efforts. GAO-16-413T. Washington, D.C.: February 11, 2016.

Air Travel and Communicable Diseases: Comprehensive Federal Plan Needed for U.S. Aviation System's Preparedness. GAO-16-127. Washington, D.C.: December 16, 2015.

Emerging Animal Diseases: Actions Needed to Better Position USDA to Address Future Risks. GAO-16-132. Washington, D.C.: December 15, 2015.

Climate Change: HHS Could Take Further Steps to Enhance Understanding of Public Health Risks. GAO-16-122. Washington, D.C.: October 5, 2015.

Biosurveillance: Challenges and Options for the National Biosurveillance Integration Center. GAO-15-793. Washington, D.C.: September 24, 2015.

Biosurveillance: Additional Planning, Oversight, and Coordination Needed to Enhance National Capability. GAO-15-664T. Washington, D.C.: July 8, 2015.

Federal Veterinarians: Efforts Needed to Improve Workforce Planning. GAO-15-495. Washington, D.C.: May 26, 2015.

Biological Defense: DOD Has Strengthened Coordination on Medical Countermeasures but Can Improve Its Process for Threat Prioritization. GAO-14-442. Washington, D.C.: May 15, 2014.

National Preparedness: HHS Has Funded Flexible Manufacturing Activities for Medical Countermeasures, but It Is Too Soon to Assess Their Effect. GAO-14-329. Washington, D.C.: March 31, 2014.

National Preparedness: HHS Is Monitoring the Progress of Its Medical Countermeasure Efforts but Has Not Provided Previously Recommended Spending Estimates. GAO-14-90. Washington, D.C.: December 27, 2013.

Homeland Security: An Overall Strategy Is Needed to Strengthen Disease Surveillance in Livestock and Poultry. GAO-13-424. Washington, D.C.: May 21, 2013.

National Preparedness: Efforts to Address the Medical Needs of Children in a Chemical, Biological, Radiological, or Nuclear Incident. GAO-13-438. Washington, D.C.: April 30, 2013.

Influenza: Progress Made in Responding to Seasonal and Pandemic Outbreaks. GAO-13-374T. Washington, D.C.: February 13, 2013.

Homeland Security: Agriculture Inspection Program Has Made Some Improvements, but Management Challenges Persist. GAO-12-885. Washington, D.C.: September 27, 2012.

Biosurveillance: Nonfederal Capabilities Should Be Considered in Creating a National Biosurveillance Strategy. GAO-12-55. Washington, D.C.: October 31, 2011.

National Preparedness: Improvements Needed for Acquiring Medical Countermeasures to Threats from Terrorism and Other Sources. GAO-12-121. Washington, D.C: October 26, 2011.

Homeland Security: Challenges for the Food and Agriculture Sector in Responding to Potential Terrorist Attacks and Natural Disasters. GAO-11-946T. Washington, D.C.: September 13, 2011.

Homeland Security: Actions Needed to Improve Response to Potential Terrorist Attacks and Natural Disasters Affecting Food and Agriculture. GAO-11-652. Washington, D.C.: August 19, 2011.

Influenza Vaccine: Federal Investments in Alternative Technologies and Challenges to Development and Licensure. GAO-11-435. Washington, D.C.: June 27, 2011.

Influenza Pandemic: Lessons from the H1N1 Pandemic Should Be Incorporated into Future Planning. GAO-11-632. Washington, D.C: June 27, 2011.

Live Animal Imports: Agencies Need Better Collaboration to Reduce the Risk of Animal-Related Diseases. GAO-11-9. Washington, D.C.: November 8, 2010.

Biosurveillance: Efforts to Develop a National Biosurveillance Capability Need a National Strategy and a Designated Leader. GAO-10-645. Washington, D.C.: June 30, 2010.

Veterinarian Workforce: The Federal Government Lacks a Comprehensive Understanding of Its Capacity to Protect Animal and Public Health. GAO-09-424T. Washington, D.C.: February 26, 2009.

Influenza Pandemic: Sustaining Focus on the Nation's Planning and Preparedness Efforts. GAO-09-334. Washington, D.C: February 26, 2009.

Veterinarian Workforce: Actions Are Needed to Ensure Sufficient Capacity for Protecting Public and Animal Health. GAO-09-178. ashington, D.C.: February 4, 2009.

Influenza Pandemic: HHS Needs to Continue Its Actions and Finalize Guidance for Pharmaceutical Interventions. GAO-08-671. Washington, D.C.: September 30, 2008.

Emergency Preparedness: States Are Planning for Medical Surge, but Could Benefit from Shared Guidance for Allocating Scarce Medical Resources. GAO-08-668. Washington, D.C.: June 13, 2008.

Influenza Pandemic: Efforts Underway to Address Constraints on Using Antivirals and Vaccines to Forestall a Pandemic. GAO-08-92. Washington, D.C.: December 21, 2007.

Influenza Vaccine: Issues Related to Production, Distribution, and Public Health Messages. GAO-08-27. Washington, D.C: October 31, 2007.

Global Health: U.S. Agencies Support Programs to Build Overseas Capacity for Infectious Disease Surveillance. GAO-08-138T. Washington, D.C.: October 4, 2007.

Agricultural Quarantine Inspection Program: Management Problems May Increase Vulnerability of U.S. Agriculture to Foreign Pests and Diseases. GAO-08-96T. Washington, D.C.: October 3, 2007.

Global Health: U.S. Agencies Support Programs to Build Overseas Capacity for Infectious Disease Surveillance. GAO-07-1186. Washington, D.C.: September 27, 2007.

Influenza Pandemic: DOD Combatant Commands' Preparedness Efforts Could Benefit from More Clearly Defined Roles, Resources, and Risk Mitigation. GAO-07-696. Washington, D.C.: June 20, 2007.

Influenza Pandemic: Efforts to Forestall Onset Are Under Way; Identifying Countries at Greatest Risk Entails Challenges. GAO-07-604. Washington, D.C.: June 20, 2007.

Avian Influenza: USDA Has Taken Important Steps to Prepare for Outbreaks, but Better Planning Could Improve Response. GAO-07-652. Washington, D.C.: June 11, 2007.

Influenza Pandemic: DOD Has Taken Important Actions to Prepare, but Accountability, Funding, and Communications Need to be Clearer and Focused Departmentwide. GAO-06-1042. Washington, D.C.: September 21, 2006.

Homeland Security: Management and Coordination Problems Increase the Vulnerability of U.S. Agriculture to Foreign Pests and Disease. GAO-06-644. Washington, D.C.: May 19, 2006.

Influenza Vaccine: Shortages in 2004-05 Season Underscore Need for Better Preparation. GAO-05-984. Washington, D.C.: September 30, 2005.

End Notes

[1] Microcephaly is a medical condition in which the circumference of the head is smaller than normal because the brain has not developed properly or has stopped growing. Guillain-Barré syndrome is a disorder in which the body's immune system attacks part of the peripheral nervous system, which can cause muscle weakness, and sometimes, paralysis.

[2] Emerging infectious diseases are infections that have newly appeared in a population or have existed but are rapidly increasing in incidence or geographic range. See S. S. Morse, "Factors in the Emergence of Infectious Diseases," *Emerging Infectious Diseases*, vol. 1, no. 1, (1995).

[3] Institute of Medicine, *Emerging Viral Diseases: The One Health Connection* (Washington, D.C.: National Academies Press, 2015).

[4] Epidemiology is the study of the distribution and determinants of diseases and injuries in human populations.

[5] Flaviviruses are positive, single-stranded, enveloped RNA viruses found in arthropods, (primarily ticks and mosquitoes), and can occasionally infect humans and other vertebrates.

[6] D.E. Lilienfeld and D. Schneider, *Lilienfeld's Foundations of Epidemiology*, 4th ed. (New York, NY: Oxford University Press, 2015).

[7] Department of Health and Human Services, Office of the Assistant Secretary for Preparedness and Response, National Health Security Strategy and Implementation Plan for 2015-2018 (Washington, D.C.: 2015).

[8] R. S. Lanciotti, et al., "Genetic and Serologic Properties of Zika Virus Associated with an Epidemic, Yap State, Micronesia, 2007," *Emerging Infectious Diseases*, vol. 14, no. 8 (2009).

[9] The Nationally Notifiable Diseases Surveillance System is a nationwide collaboration that enables public health at the local, state, territorial, federal, and international levels to share notifiable disease-related health information and uses this information to monitor, control, and prevent the occurrence and spread of those diseases.

[10] D. Musso et al., "Potential Sexual Transmission of Zika Virus," *Emerging Infectious Diseases*, vol. 21, no. 2 (2015).

[11] Zika virus components have been detected in semen for up to 62 days post initial infection, but the duration of viable Zika virus transmission via semen is not known.

[12] U.S. Food and Drug Administration, "FDA Issues Recommendations to Reduce the Risk for Zika Virus Blood Transmissions in the United States," Released February 16, 2016, Accessed February 25, 2016, http://www.fda.gov/NewsEvents/Newsroom/PressAnnouncements/ucm486359.htm.

[13] C. G. Victoria, "Microcephaly in Brazil: how to interpret reported numbers?" *The Lancet*, vol. 387, no. 10019, (2016).

[14] J. Mlakar, et al., "Zika Virus Associated with Microcephaly," *New England Journal of Medicine*, (2016).

[15] European Centre for Disease Prevention and Control. *Rapid risk assessment: Zika virus infection outbreak, French Polynesia.* (Stockholm: February 2014).

[16] Serology refers to the diagnostic identification of antibodies in the serum of blood.

[17] The Council of State and Territorial Epidemiologists is an organization of member states and territories representing public health epidemiologists. Every year, their position statements are used to update the case definitions of nationally notifiable diseases, and the organization encourages state health departments to report cases of these selected diseases to CDC's National Notifiable Diseases Surveillance System.

[18] In a comment published in The Lancet, Victora, C.G. et al., reported that "Before 2015, the annual numbers of reported cases of microcephaly in Brazil were consistently below 200. Between mid-2015 and Jan 30, 2016, 4783 suspected cases of microcephaly were reported, including newborn and fetal losses. Of these, 1103 cases have completed clinical, laboratory, and imaging examinations, and 404 (36.2%) were classified as confirmed cases of microcephaly. Among the confirmed cases, brain abnormalities were detected by imaging in 387 babies and Zika virus was detected in 17 babies, including in two fetal losses. The remaining 709 cases were discarded and 3670 suspected cases of microcephaly remain under investigation."

[19] D. L. Heymann, et al., "Zika virus and microcephaly: why is this situation a PHEIC?" *The Lancet*, vol. 387, no. 10020, (2016).

[20] Reverse transcription polymerase chain reaction (RT-PCR) is a technique to detect or quantify RNA, according to a scientific journal. It uses a reverse transcriptase enzyme to convert RNA to DNA, followed by PCR to amplify the DNA. An immunoglobulin test measures the level of certain immunoglobulins, or antibodies, in the blood. CDC defines antibodies as proteins found in the blood that are produced in response to foreign substances (e.g., bacteria or viruses) invading the body. Antibodies protect the body from disease by binding to these organisms and destroying them.

The Plaque Reduction and Neutralization Test (PRNT) measures the amount of the neutralizing antibodies in the serum of the infected individual and determines the level of protective antibodies this individual has towards the infecting virus. This test is based on how viruses and antibodies interact, resulting in inactivation of virus – when the correct antibody is present - such that it is no longer able to infect and replicate in living cells.

[21] Specificity is the percentage of persons who do not have a given condition who are identified by the assay as negative for the condition.

[22] A. B. Nuss et al., "Dopamine Receptor Antagonists as New Mode-of-Action Insecticide Leads for Control of *Aedes* and *Culex* Mosquito Vectors," *PLOS Neglected Tropical Diseases*, vol. 9, no. 3, (2015).

[23] N. L. Achee, et al., "A Critical Assessment of Vector Control for Dengue Prevention," *PLOS Neglected Tropical Diseases,* vol. 9, no. 5, (2015).

[24] L. Alphey, "Genetic control of mosquitoes," *Annual Review of Entomology*, vol. 59 (2013).

[25] NIH also provided information about additional priority research areas: "Notice of NIAID's Interest to Highlight High-Priority Zika virus (ZIKV) Research Areas" and "NHLBI Announces High Priority Interest in Zika Virus-related Blood Supply Safety and Transfusion Research".

[26] U.S. Centers for Disease Control and Prevention, Morbidity and Mortality Weekly Report, *Interim Guidelines for Pregnant Women During a Zika Virus Outbreak — United States, 2016*, (Washington, D.C.: January 22 2016).

[27] According to scientific literature, a case-control study is designed to help determine if an exposure (in this case, a viral infection) is associated with an outcome, such as a disease or condition (such as microcephaly or Guillain-Barré syndrome). This is done by identifying the cases (a group known to have the outcome) and the controls (a group known to be free of the outcome). By looking back in time to learn which subjects in each group had the exposure(s), a researcher can compare the frequency of the exposure in the case group to the control group and hypothesize a link between the exposure and condition. From: S. Lewallen and P. Courtright, "Epidemiology in practice: case-control studies," *Community Eye Health,* vol. 11, no. 28, (1998).

[28] B. O. Saad and R. H. Beigi, "Pregnancy in the Time of Zika: Addressing Barriers for Developing Vaccines and Other Measures for Pregnant Women," *The Journal of the American Medical Association*, (2016).

In: The Zika Virus
Editor: Veronica Salazar

ISBN: 978-1-53610-124-9

Chapter 5

TESTIMONY OF ANTHONY S. FAUCI, DIRECTOR, NATIONAL INSTITUTE OF ALLERGY AND INFECTIOUS DISEASES, NATIONAL INSTITUTES OF HEALTH. HEARING ON "EXAMINING THE U.S. PUBLIC HEALTH RESPONSE TO THE ZIKA VIRUS"*

Mr. Chairman, Ranking Member DeGette, and Members of the Subcommittee:

Thank you for the opportunity to discuss the National Institutes of Health (NIH) research response to Zika virus, an emerging public health threat of international concern. I direct the National Institute of Allergy and Infectious Diseases (NIAID), the lead NIH institute for conducting and supporting research on emerging and re-emerging infectious diseases, including those caused by flaviviruses such as Zika virus.

The Administration is taking appropriate action to protect the American people and, as you know, it announced a request to Congress for approximately $1.9 billion in emergency funding to enhance ongoing efforts to prepare for and respond to outbreaks of the Zika virus, both domestically and internationally. This includes funding for work on the development of

* This is an edited, reformatted and augmented version of testimony presented March 2, 2016 before the House Committee on Energy and Commerce, Subcommittee on Oversight and Investigations.

vaccines and diagnostics and to improve scientific understanding of the disease.

The overarching mission of NIAID is to conduct and support research to better understand, treat, and prevent infectious and immunologic diseases. This is accomplished through a spectrum of research, from basic studies of the mechanisms of disease to applied research focused on developing interventions such as diagnostics, therapeutics, and vaccines. As part of this mission, NIAID has a dual mandate encompassing both research on ongoing public health issues and the capability to respond rapidly to newly emerging and re-emerging infections such as Zika virus.

These emerging and re-emerging disease threats, whether man-made or naturally occurring, are perpetual challenges, in part due to the capacity of microbial pathogens to evolve rapidly and adapt to new ecological niches. To address the challenges posed by emerging infectious diseases, NIAID employs both targeted, disease-specific research as well as broad-spectrum approaches. NIAID maximizes its efforts by prioritizing the development of drugs effective against multiple bacteria or viruses, and "platform" technologies to facilitate rapid development of vaccines and diagnostics applicable to multiple infections.

NIAID is well-positioned to rapidly respond to infectious disease threats as they emerge by leveraging fundamental, basic research efforts; domestic and international research infrastructure that can be quickly mobilized; and productive partnerships with industry. NIAID provides preclinical research resources to scientists in academia and private industry worldwide to advance translational research against emerging and re-emerging infectious diseases. These resources are designed to bridge gaps in the product development pipeline and lower the scientific, technical, and financial risks incurred by industry in order to incentivize them to partner with us in the advanced development of effective countermeasures. NIAID also supports our Vaccine and Treatment Evaluation Units (VTEUs), a research network that conducts clinical trials to quickly investigate promising therapeutic and vaccine candidates when public health needs arise. NIAID collaborations with other federal agencies, including those undertaken within the Department of Health and Human Services (HHS) Public Health Emergency Medical Countermeasures Enterprise (PHEMCE), help advance progress against newly emerging public health threats. In addition, partnerships with academia, the biotechnology and pharmaceutical industries, and international researchers and organizations such as the World Health Organization (WHO) and WHO's regional office, the Pan American Health Organization (PAHO), are integral to these efforts.

OVERVIEW OF ZIKA VIRUS

Zika virus is a flavivirus. These viruses typically are transmitted by mosquitoes and often have the ability to spread quickly to new geographic locations because of the widespread prevalence of these vectors. Other well-known flaviviruses include dengue virus and yellow fever virus; like Zika virus they are transmitted by *Aedes* mosquitoes. Zika virus was discovered in monkeys in Uganda in 1947 and is now endemic to Africa and Southeast Asia. During the past decade it has emerged in other areas of the world, including Oceania, the Caribbean, and Central and South America, where countries, notably Brazil, are currently experiencing unprecedented Zika transmission.

Infections caused by Zika virus are usually asymptomatic. About 20 percent of infected individuals experience clinical symptoms such as fever, rash, joint pain, and conjunctivitis (red eyes). Symptoms of Zika virus infection in humans are typically mild and brief, with very low hospitalization and fatality rates. The recent outbreak of Zika virus disease in Brazil has coincided with a reported increase in the number of infants born with microcephaly, a birth defect characterized by an abnormally small head resulting from an underdeveloped and/or damaged brain. In addition, increases in suspected cases of Guillain-Barré syndrome (GBS), a rare, acute, immune-mediated peripheral nerve disease that leads to weakness, sometimes paralysis, and infrequently, respiratory failure and death, have been noted in Brazil and other countries in the Americas.

Further research is needed to better understand the effect of Zika virus infection on the body, particularly during pregnancy; to investigate the potential relationship between Zika infection and congenital abnormalities including microcephaly, as well as to explore the potential relationship between Zika infection and GBS; and to develop better diagnostics, vaccines and treatments, and new methods of vector control. Currently, no vaccines or specific therapeutics are available to prevent or treat Zika virus disease. Improved diagnostic tests also are needed because Zika virus infection causes non-specific symptoms or no symptoms at all and can be difficult to distinguish by antibody screening tests from other mosquito-borne infections such as dengue, malaria, and chikungunya. Moreover, current antibody screening tests can be falsely positive or inconclusive if the individual was previously infected with related viruses such as dengue, which is prevalent in South America and the Caribbean. Therefore, a positive result with the antibody screening test requires an additional test to confirm the diagnosis.

NIH Research on Zika Virus

NIAID has a longstanding commitment to flavivirus research, including extensive efforts to combat diseases such as dengue, West Nile virus, and yellow fever. This research has informed our understanding of the viral genetics, vector biology, and pathogenesis of flaviviruses and provides a strong foundation for our efforts to learn more about Zika virus. NIAID has responded to the newly emerging Zika virus disease outbreak by expanding our portfolio of basic research on Zika virus and other flaviviruses. NIAID also is accelerating efforts to develop improved diagnostics and candidate therapies for Zika virus as well as prioritizing the development of Zika virus vaccines. In addition, screening tests and pathogen reduction technologies are critically important to assure safety of the U.S. blood supply.

The emergency funding for NIH would support development of vaccines to prevent Zika virus infection, from the discovery phase through preclinical and eventually clinical testing. In addition, the funds would support basic research to understand the natural history, viral biology and pathogenesis, including potential links to microcephaly; establishment of animal models to test candidate countermeasures; development of rapid, sensitive, and specific diagnostic tests; and discovery and preclinical development of new therapeutics to treat disease caused by Zika virus. This research is necessary to better understand this emerging infection and uncover the best ways to diagnose, treat, and prevent Zika virus disease.

In January 2016, NIAID issued a notice to researchers highlighting NIH's interest in supporting research and product development to combat Zika virus. Areas of high priority include basic research to understand viral replication, pathogenesis, and transmission, as well as the biology of the mosquito vectors; potential interactions with co-infections such as dengue and yellow fever viruses; animal models of Zika virus infection; and novel vector control methods. In addition, NIH is soliciting Zika virus research to develop sensitive, specific, and rapid clinical diagnostic tests; drugs against Zika virus as well as broad spectrum therapeutics against multiple flaviviruses; and effective vaccines and vaccination strategies.

NIAID also is partnering with other NIH institutes, the *Eunice Kennedy Shriver* National Institute of Child Health and Human Development (NICHD), the National Institute of Neurological Disorders and Stroke (NINDS), and the National Institute of Dental and Craniofacial Research, to accelerate Zika virus research as it relates to the mother-infant pair. The Institutes issued a notice that indicates NIH's interest in supporting research to understand transmission, optimal screening and management in pregnancy, and the mechanisms by which Zika virus affects the developing nervous system, including potential links to microcephaly and other congenital abnormalities.

Developing Tools to Combat Zika Virus

In response to public health concerns about Zika virus, NIAID has accelerated ongoing flavivirus research efforts to speed the development of tools that could help control current and future outbreaks of Zika virus.

Vector Control

For many years, NIAID has supported extensive research to understand the biology of mosquitoes to help develop tools to limit the spread of deadly mosquito-borne diseases such as dengue and malaria. This research aids in vector control strategies to reduce mosquito bites or limit mosquito populations. In the Americas, Zika virus is transmitted primarily by *Aedes aegypti* mosquitoes, and vector control or other methods to prevent exposure to these mosquitoes are currently the only ways to prevent Zika infection. NIAID plans to support vector competence studies to test various mosquito species for their ability to carry and transmit Zika virus and for insecticide resistance. Understanding the specific mosquito species involved in Zika outbreaks and which insecticides may be effective against them will aid current vector control efforts and may inform novel mosquito control strategies in the future.

Diagnostics

Accurate diagnostic tests for Zika virus infection are needed to distinguish it from other flavivirus infections and to identify women who have been infected with Zika virus during pregnancy and may be at risk for developing fetal complications. Blood, organ, and tissue donor screening tests are also needed to assure the safety of transfusion and transplantation in areas of active mosquito-borne virus transmission. Currently, Zika virus itself can often be detected during the acute phase of infection and up to seven days after the onset of symptoms using diagnostic tests for viral RNA (RT-PCR test). While prior infection can be detected by testing for the presence of antibodies against Zika virus, assays for Zika antibodies may also detect or cross-react with antibodies against other flaviviruses, particularly dengue virus. For this reason, a positive antibody test does not definitively confirm prior Zika virus infection in the setting of possible co-infection or prior infection with dengue and other related viruses, and separate confirmatory testing is required. This is a particular concern in South America where there is a high level of exposure to other flaviviruses, particularly dengue virus.

To facilitate the development of improved Zika virus diagnostic tests, NIAID grantees are working to generate antibodies that can distinguish between Zika virus and dengue virus. They also are working to identify biosignatures unique to Zika infection that could form the basis of additional rapid, specific, and sensitive diagnostic tests. In addition, NIAID is pursuing the development of a mouse model of Zika virus infection that could be used to test new diagnostic and therapeutic tools.

Vaccines

A safe and effective Zika vaccine would be a very valuable tool to help stop the spread of infection and prevent future outbreaks. NIAID is investigating multiple Zika virus vaccine candidates, including vaccines based on technologies that have shown promise in targeting other flaviviruses. The NIAID Vaccine Research Center (VRC) is pursuing a DNA-based vaccine for Zika virus that is similar to a West Nile virus vaccine previously developed by NIAID. The West Nile vaccine candidate was shown in Phase 1 testing to be safe and generated a strong immune response in humans, offering a model for Zika vaccine development. NIAID scientists also are designing a live, attenuated vaccine, using an approach similar to that used for making a vaccine against the closely related dengue virus. The dengue vaccine candidate showed an excellent safety profile and generated strong immune responses in early-phase clinical trials. In January, a large Phase 3 trial assessing the dengue vaccine candidate was launched in Brazil in collaboration with the

Butantan Institute. In addition, NIAID grantees are in the early stages of developing a Zika virus vaccine based on a recombinant vesicular stomatitis virus – the same animal virus used successfully to create an investigational Ebola vaccine. Plans are underway to evaluate this potential vaccine construct in tissue culture and animal models.

While these approaches are promising, it is important to realize that the development of investigational vaccines and the clinical testing to establish whether they are safe and effective takes time. Although a safe and effective, fully licensed Zika vaccine will likely not be available for a few years, we plan to begin early-stage clinical testing of one or more NIAID-supported vaccine candidates in 2016.

Therapeutics

NIAID has an active program to screen for antiviral drugs active against viruses in the flavivirus family, including dengue, West Nile, yellow fever, and Japanese encephalitis viruses, as well as the closely related hepatitis C virus. NIAID has enhanced these efforts with the recent development of an assay to test compounds for antiviral activity against Zika virus. NIAID will make this test available to the research community and will soon test 10 antiviral compounds with activity against other flaviviruses to determine if they are effective against Zika virus. Promising drug candidates identified by the assay could be further tested in a small animal model of Zika virus infection developed with NIAID support. The ultimate goal of NIAID-supported flavivirus therapeutic research is to develop a broad-spectrum antiviral drug that could be used against a variety of flaviviruses, including Zika.

Emergency Request for Vaccine Research and Diagnostic Development and Procurement

As I noted in the introduction to my testimony, the Administration has announced an emergency funding request of approximately $1.9 billion to combat the Zika virus both domestically and internationally. Included in the request are resources for Zika-related vaccine research, rapid advanced development, and commercialization of new vaccines and diagnostic tests for Zika virus. The funding will allow NIH to build upon existing resources and work to develop a vaccine for Zika virus and the chikungunya virus, which is spread by the same type of mosquito. Funding will accelerate this work and improve scientific understanding of the disease to inform the development of additional tools to combat it. The request also includes resources for FDA to support Zika virus medical product development, including the next-generation diagnostic devices. We look forward to working with the Congress to implement this request.

Collaborations

Investigation of emerging and re-emerging infectious diseases requires expertise from a variety of fields. In the case of Zika virus, studies of virology, immunology, natural history, neurology, and neonatology will be required to fully understand the pathogenesis of this infection. As mentioned previously, NIAID is partnering with other NIH institutes including NICHD and NINDS to better understand the potential association between Zika virus infection and neonatal defects, particularly microcephaly.

NIAID also is employing partnerships with research institutions in South America to advance research on Zika virus infection; additional collaborations with academic, industry, and government partners are under active exploration. NIAID held a joint meeting in December 2015 with Brazilian research institute Fiocruz in which Zika was a key area of concentration. In addition, NIAID is collaborating with other HHS agencies in responding to the Zika epidemic. For example, NIAID, CDC, BARDA, ASPR, and FDA are jointly convening a Zika virus workshop on March 28-29, 2016, where the latest information on Zika virus will be discussed by experts from Federal Agencies, academia, and pharmaceutical and biotechnology companies. Topics to be addressed at the workshop include virology, epidemiology, possible links to microcephaly, and efforts to develop diagnostics, therapeutics, and vaccines.

Conclusion

NIH is committed to continued collaboration with HHS agencies and other partners across the U.S. government in advancing research to address Zika virus infection, and we look forward to working with the Congress to implement the President's emergency funding request. As part of its mission to respond rapidly to emerging and re-emerging infectious diseases throughout the world, NIAID is expanding our efforts to elucidate the biology of Zika virus and employ this knowledge to develop needed tools to diagnose, treat, and prevent disease caused by this virus. In particular, NIAID will pursue the development of safe, effective vaccines to prevent disease caused by Zika and chikungunya viruses.

In: The Zika Virus
Editor: Veronica Salazar

ISBN: 978-1-53610-124-9

Chapter 6

TESTIMONY OF JOSEPH M. CONLON, TECHNICAL ADVISOR, AMERICAN MOSQUITO CONTROL ASSOCIATION. HEARING ON "EXAMINING THE U.S. PUBLIC HEALTH RESPONSE TO THE ZIKA VIRUS"*

Good morning, Mr. Chairman and Members of the Subcommittee. My name is Joseph Conlon and I am a retired U.S. Navy medical entomologist serving as Technical Advisor for the American Mosquito Control Association (AMCA), a nonprofit organization of 1600 members dedicated to enhancing health and quality of life through the suppression of mosquitoes and other vectors of public health importance. I welcome this opportunity to provide a mosquito control perspective to the deliberations of this committee concerning Zika Virus and will limit my testimony to mosquito management methodologies that contribute to its control.

The introduction and spread of Zika Virus in Central and South America and the Caribbean in 2014 reawakened in the United States an appreciation of mosquitoes as vectors of diseases. I use the term "reawakened" advisedly, for mosquito-borne diseases such as malaria and yellow fever were once quite prevalent in the United States and, indeed, played a major part in shaping our nation's destiny. These diseases no longer claim victims in the United States as a matter of course largely due to the exemplary efforts of organized

* This is an edited, reformatted and augmented version of testimony presented March 2, 2016 before the House Committee on Energy and Commerce, Subcommittee on Oversight and Investigations.

mosquito control agencies, in conjunction with an enlightened and effective public health infrastructure. But other mosquito-borne diseases are on the horizon and public health agencies need to be prepared to meet the challenges they present.

Progress continues in defining the transmission dynamics of Zika Virus so as to allow for its prevention and control. However, considering that it is a comparatively recent epidemiological phenomenon, there remains much to learn in order to establish and verify baseline data. The virus is thought to be primarily transmitted from human to human by the bite of infected *Aedes aegypti* and (possibly) *Aedes albopictus* species of mosquitoes. Of these, *Aedes aegypti* has been primarily responsible for transmitting the disease due to its preference to bite humans both day and night and its predilection for biting the lower extremities. Moreover, females frequently take multiple partial bloodmeals, often from different individual humans, not only increasing the likelihood of feeding on an infectious human but also leading to single infectious females potentially feeding on and infecting multiple humans within a relatively short time period.

Aedes aegypti and *Aedes albopictus* are notoriously difficult to prevent or kill. They live inside our houses under furniture, beds and in closets; their eggs can withstand months of drying, and their young can develop in water containers as small as a bottle cap. Virtually any collection of stagnant water in containers, treeholes, leaf axils, etc. can serve as egg-laying habitat for these species. Thus, draining wet areas doesn't prevent their development around our homes and yards; predators do not consume enough to provide effective control; and repellents, while protecting individuals, just redirect the adult mosquitoes to bite somebody else.

Aedes aegypti has been found in isolated areas of California, and from Arizona, New Mexico, Texas, Louisiana, Mississippi, Alabama, Florida, Georgia and South Carolina. *Aedes albopictus* is a more cold-hardy species and its range includes those states but extends northward to Illinois and New York. Its range is thought to be defined by an average winter temperature of 50 degrees Fahrenheit.

MOSQUITO CONTROL AS CURRENTLY PRACTICED IN THE UNITED STATES

The first mosquito control districts were established in NJ in 1912. California and Florida followed suit in 1915 and 1925, respectively. In the ensuing years, mosquito control districts and state agencies were established nationwide so that there are now more than 700 recognized districts within the United States along with over 1000 control entities within municipal service programs.

The integrated mosquito management methods currently employed by these organized control agencies and endorsed by the CDC and EPA are comprehensive and specifically tailored to safely counter each stage of the mosquito life cycle. Larval control through water management and source reduction, where compatible with other land management uses, is a prudent pest management alternative - as is use of the environmentally friendly EPA-approved larvicides currently available. When source elimination or larval control measures are clearly inadequate, or in the case of imminent disease, the EPA and CDC have emphasized in a published joint statement the need for considered application of adulticides by certified applicators trained in the special handling characteristics of these products.

In 2009 the AMCA published its "Best Management Practices for Integrated Mosquito Management" and promulgated it to its membership as a set of guidelines for safe, effective and efficient vector control programs. The AMCA is currently in the process of developing guidelines specifically geared toward control of *Aedes aegypti* and *Aedes albopictus*. This will emphasize urban mosquito control and the paradigm shift in control practices from floodwater/salt marsh mosquitoes (predictable habitats) to container-inhabiting mosquitoes (unpredictable and ubiquitous in the peridomestic environment). In addition, the unique man-power, resources, and funding needed for proactive urban control will be covered. AMCA hopes to have a finalized version available as early as possible in 2016/2017 timeframe.

States thought to be particularly susceptible to Zika introduction, such as Florida California, and Texas, have excellent integrated mosquito control programs in many areas, but emergency funds to respond to a Zika outbreak are needed if timely interventions to contain outbreaks in these districts and elsewhere are to be realized. Critically, many potential ports of entry throughout the U.S. have limited or no mosquito control available and there is a pressing need for funding to establish sustainable mosquito control programs in these areas.

Components of successful mosquito-borne disease control programs as currently practiced in the United States include, but are not limited, to the following:

Prevention

Surveillance - A sustained, consistent surveillance program targets the particular vector species, mapping their larval habitats by season, and documents the need for control through larval and adult trapping regimens. It thus also monitors the effectiveness of the control program. Appropriate and timely response to surveillance data is the key to preventing human disease associated with Zika. Control activity should then be intensified in response to evidence of virus in imported cases, as deemed necessary by the local statutory authority.

Public Information and Outreach – Studies have shown that information programs, while crucial to the overall prevention/control strategy, only exert a moderate effect on modifying population behaviors related to personal protective measures. Effective programs include development of a community task force, interventions to improve access to window screening materials or repellents, and social marketing to reinforce preventive behaviors. These are critical components of any mosquito control program, but cannot, in and of themselves, replace established prevention/control methodologies.

Source Reduction - Removing breeding habitat is the most effective long-term mosquito preventative/control and includes activities as simple as the proper disposal of used tires, paint cans and trash, in addition to the cleaning of rain gutters and bird baths, by individual property owners.

Control

Reducing the contact between the vector mosquito and humans - This is accomplished through removing, modifying or treating larval habitats; modification or removal of adult mosquito resting areas, adulticide treatments when indicated, and use of repellents. All of these interventions can be problematic regrding control of *Aedes aegypti*.

Best Management Practices (BMP) - Most larger mosquito control districts in states having potential ports of entry, such as California, Texas and Florida, usually employ a phased response based upon mosquito trapping data. Such programs consist of public education emphasizing personal protection and residential source reduction; municipal larval control to prevent repopulation of the area with competent vectors; adult mosquito control to decrease the density of infected, adult mosquitoes in the area; and continued surveillance to monitor virus activity and efficacy of control measures.

How Mosquito Control Can Be Improved

- Mosquito surveillance tools for adult *Aedes aegypti* are ineffective for small populations and need to be improved and made more widely available to marginally funded programs.
- Training in the use and interpretation of survey results should be made available to both program managers and public health officials as continuing education adjuncts.
- Additional training resources and comprehensive participation in workshops and professional mosquito control conferences are key components of program upgrades.
- Faster and more efficient communications between public health laboratories and vector control programs need to be explored and implemented. The speed with which the public health sector can respond to either imported or locally-acquired cases will ultimately depend on the turnaround time between testing of suspect cases, laboratory confirmation and promulgation of surveillance results. In addition, variable interpretations of the Health Insurance Portability and Accountability Act (HIPAA) may delay release of results, further restricting timely interventions by vector control agencies. Mechanisms need to be developed that require health departments at all levels to disclose to local vector control agencies the location of suspect cases of Zika while not violating HIPAA provisions. This will ensure more timely and targeted initiation of intervention measures that will prevent or rapidly contain outbreaks.
- Materials and training regarding pesticide resistance testing should be implemented nationwide.

- o Funds should be made available to underwrite new data call-ins that might influence a pesticide registrant's decision to keep products on the market
- o New chemical, genetic, and biological control methodologies and means of dispersal should be developed and field validated.
- o Conventional pesticides are losing ground to resistance. New tools are in development (spatial repellents, ATSB, lethal ovitraps, intracellular biological control (*Wolbachia*), RNAi, GMO mosquitoes, and new pesticide chemical classes require more field validation prior to widespread use.
- o Rapid diagnostic tests for Zika and other arboviral agents need development and deployment to mosquito control entities to provide timely notification of infected mosquitoes prior to human cases.

THE ROLE OF FEDERAL GOVERNMENT

The federal government has a crucial role to play in how vector control is accomplished. Several agencies either conduct (e.g., USDA and to a lesser degree CDC) research on applied vector control topics, or fund (NIH) projects (e.g., *Wolbachia*) into novel technologies, FEMA supports large-scale vector control in declared disasters (under Stafford Act stipulations).

In addition, the Centers for Disease Control and Prevention (CDC) Division of Vector-borne Disease provides national expertise in mosquito-borne disease prevention and control in concert with members of the AMCA. They have also served as an effective conduit via the Expanded Laboratory Capacity (ELC) program for funding that can increase arboviral testing capacities at both state and local levels. To this end, we encourage Congress to authorize supplemental funding for 2016 and increased funding in FY 2017 to support vector-borne disease surveillance. Funding should be allocated through the Centers for Disease Control and Prevention's (CDC's) Division of Vector-Borne Diseases for dispersal to localities most vulnerable to disease introduction while lacking the capabilities.

The Mosquito Abatement for Public Health Act (MASH), Public Law 108-75, was passed during the first session of the 108th Congress but has not received appropriations for its implementation. This Act is a mechanism by which local governments could receive matching federal funds up to \$100,000 for the establishment and/or enhancement of a mosquito abatement program through the CDC. The MASH Act authorized \$100,000,000 in funds for FY 2003, but, as of this hearing, Congress has still not appropriated any funds to cover its provisions. In addition, it requires NIEHS to conduct or support research into methods to control the population of insects and vermin that transmit dangerous diseases to humans. The AMCA fully supports the MASH Act and requests action to appropriate the funds for its full implementation to bolster long-term vector surveillance and control.

The critical challenge for the nation's public health system will be modifying the behavior of the US population to eliminate oviposition sites for these species. Vector control agencies do not possess the resources to provide the sustained service for each resident in their jurisdiction that is required to successfully control invasive *Aedes*. To accomplish that end via source reduction will entail a national effort, along the lines of the "buckle up" seat belt, Smoky the Bear, and "don't be a litterbug" campaigns of the 1960s. A sustained national program will required continued funding into the foreseeable future if the benefits from reduced introduction and spread of exotic mosquito-borne diseases are to be realized.

We further see a federal government role in maintaining our present, proven chemical tools at risk of losing their registration. In 1996 Congress unanimously approved the Food Quality Protection Act (FQPA) (PL 104-170) to modernize the regulation of pesticides and expand data requirements to demonstrate their safety to people and the environment. A key element was authorization to use federal funds when the cost of new data for public health pesticides was more than their producers could afford, putting registration at risk. Unfortunately, these essential funds have never been appropriated, and we are now losing critical public health tools because the cost to prove their safety is higher than return on investment.

CONCLUSION

Vector-borne diseases, whether ancient like malaria or relatively new like Zika, are an unfortunate reality, and Zika won't be the last to challenge our vector-borne disease control capabilities. There are many factors that contribute to the emergence of novel vector-borne diseases, including poverty, climate change, and global trade. They will require long-term solutions.

Mosquito control capabilities in the United States are the finest in the world, but in many cases, fall short of providing the level of protection needed to prevent outbreaks of mosquito-borne disease in all potential ports of entry. Increased tourism and trade make continual introductions of exotic diseases inevitable and resources for sustainable prevention and control programs must be made available so that future imported cases of exotic diseases beyond Zika can be contained and eliminated before their establishment and spread. Our citizenry deserves no less.

SUMMARY

- Capacities to address mosquito-borne diseases in the U.S. are better than in the past, but require significant improvement to successfully meet the mosquito-borne disease threats such as Zika Virus surely to come our way in the future.
- Increased capacity must include capabilities to address survey control strategies applicable to the widest possible variety of vector-borne disease transmission cycles.
- The unique bionomics of Aedes aegypti require better surveillance and prevention/control tools and strategies over the long-term.
- The Federal role should consist of exploring and implementing funding mechanisms devoted to increasing sustainable integrated vector surveillance and control capabilities in areas with organized mosquito control and to develop startup programs in areas of potential risk.
- The MASH act proposes just that sort of thing – providing support for developing new vector control where lacking, and enhancing capacities where they exist but are inadequate.
- Fully fund the Food Quality Protection Act provisions to use federal funds when the cost of new data for existing public health pesticides is more than their producers can afford, putting their registration at risk.
- Increase the CDC/DVBD annual budget to levels that allow funds to be allocated directly for testing of mosquito pools conducted at the local mosquito control program level or intrastate testing collaborative below the state level.

In: The Zika Virus
Editor: Veronica Salazar

ISBN: 978-1-53610-124-9

Chapter 7

STATEMENT OF DR. LUCIANA BORIO, ACTING CHIEF SCIENTIST, U.S. FOOD AND DRUG ADMINISTRATION. HEARING ON "EXAMINING THE U.S. PUBLIC HEALTH RESPONSE TO THE ZIKA VIRUS"*

INTRODUCTION

Good morning Chairman Murphy, Ranking Member DeGette, and members of the Subcommittee. I am Dr. Luciana Borio, Acting Chief Scientist at the Food and Drug Administration (FDA or the Agency). Thank you for the opportunity to appear today to discuss FDA's actions in response to the Zika virus outbreak.

As you know, Zika virus was first identified in 1947 in Uganda. Since then, sporadic cases and a few outbreaks have been recognized in a number of locations, including parts of Africa, Asia, and the Pacific. However, the situation has changed dramatically since May 2015, when the first local transmission of Zika virus in the Americas was confirmed in Brazil. By December 2015, Brazil had reported more than 56,000 suspected cases of Zika virus and has since stopped counting cases due to the magnitude of the

* This is an edited, reformatted and augmented version of a statement presented March 2, 2016 before the House Committee on Energy and Commerce, Subcommittee on Oversight and Investigations.

outbreak, which Brazilian national authorities estimate to encompass from approximately 500,000 to 1.5 million cases.

Zika virus is now spreading throughout Latin America and the Caribbean, as well as in parts of the Pacific and Africa. As of February 25, 2016, 34 countries and territories, including the Commonwealth of Puerto Rico, the U.S. Virgin Islands, and American Samoa, have active, local transmission of Zika virus.

Potential links between Zika virus infection and neurological complications (e.g., Guillain-Barré Syndrome), as well as microcephaly and other poor pregnancy outcomes in babies of mothers who were infected with Zika virus during their pregnancy have dramatically increased concerns about the risks associated with Zika virus. On February 1, 2016, the World Health Organization (WHO) declared the recent clusters of microcephaly and other neurological disorders a public health emergency of international concern.

No locally transmitted mosquito-borne Zika virus disease cases have been reported in the continental United States to date. There have, however, been travel-associated cases, as well as local cases of sexual transmission to women whose only known risk factor was sexual contact with an infected male partner who traveled to an area with active transmission. These imported cases have the potential to result in local spread of Zika virus in areas of the United States where *Aedes* mosquitoes that transmit Zika virus are found. CDC and state, local, and territorial health departments are monitoring for any such events.

FDA Response to the Zika Virus Outbreak

FDA is working in collaboration with other components of the Department of Health and Human Services (HHS), including the Office of the Assistant Secretary for Preparedness and Response (ASPR) and its Biomedical Advanced Research and Development Authority (BARDA), the National Institutes of Health (NIH), and the Centers for Disease Control and Prevention (CDC), as well as with partners across the U.S. Government and with international partners to respond to the Zika virus outbreak. FDA's primary areas of activity include: (1) protecting the safety of the nation's blood supply and tissues for transplantation; (2) facilitating the development of donor screening and medical diagnostic tests that may be useful for identifying the presence of, or prior exposure to, Zika virus; (3) supporting the development of investigational vaccines and therapies; (4) reviewing proposals for the use of innovative strategies to help suppress the population of virus-carrying mosquitoes; and (5) protecting the public from fraudulent products that claim to prevent, diagnose, treat, or cure Zika virus disease.

Blood Supply and Tissue Safety

In the absence of an available donor screening assay, the risk of transmission of Zika virus by blood transfusion is considered likely, based on the most current scientific evidence of how Zika virus and other flaviviruses (including West Nile and dengue) are spread, as well as recent reports of transfusion-associated infection outside of the United States. Realizing this risk, and to better protect the U.S. blood supply, FDA issued new Zika-related blood donor guidance. In areas with active Zika virus transmission, FDA recommends that whole blood and blood components for transfusion be obtained from areas of the United States without active transmission. Blood establishments may continue collecting and preparing platelets and plasma if an FDA-approved, pathogen-reduction device is used. FDA also recommends deferral from donating blood for four weeks for individuals in areas without active Zika transmission if they a) have been to areas with active Zika virus transmission; b) have recovered from an illness with symptoms suggestive of Zika virus infection that developed within two weeks of exposure in an affected area; or c) have had sexual contact with a man who has traveled to, or resided in, an area with active Zika virus transmission during the prior three months. The guidance also recommends that blood establishments update donor education materials with information about Zika virus signs and symptoms, ask potentially affected donors to refrain from giving blood, and update the donor history questionnaire to elicit information on recent exposure to an affected area. FDA is working with public health authorities in territories with confirmed Zika virus to take rapid and appropriate steps to ensure that safe blood is available.

In addition, to advance blood safety against Zika virus, FDA is facilitating the development of screening tests for identifying the presence of the virus in donated blood and pathogen-reduction technologies to kill viruses in blood components through establishment of reference materials, clarification of regulatory pathways, and interactive review of sponsor-submitted data.

Furthermore, FDA is developing guidance that will address appropriate donor deferral measures for human cells, tissues, and cellular and tissue-based products. This guidance is particularly important given evidence of sexual transmission of Zika virus.

Diagnostic Tests

Unfortunately, there are no commercially available diagnostic tests cleared by FDA for the detection of Zika virus. Two types of diagnostic tests are needed for Zika virus: (1) a test to diagnose acute infection; and (2) a test to assess whether individuals, especially pregnant women, who were potentially exposed to Zika virus were actually infected. In keeping with FDA's practice for responding to emerging infectious disease outbreaks, FDA has been reaching out to potential diagnostic manufacturers to encourage them to develop needed diagnostic tests for Zika virus. In addition, FDA is working interactively with diagnostic manufacturers interested in developing diagnostic tests for Zika virus to help accelerate development programs, including clarifying data requirements for the authorization of the use of Zika diagnostic tests under FDA's Emergency Use Authorization authority (EUA). FDA has authorized the use of a Zika virus diagnostic test developed by CDC under EUA for the qualitative detection of Zika virus-specific immunoglobulin M (IgM) antibodies by qualified laboratories. This diagnostic test can help expand domestic readiness for Zika virus by enabling the identification of patients recently infected with Zika virus in support of response efforts. The authorized test has already been made available to 19 states and territories. FDA is ready to authorize the use of additional

Zika virus diagnostic tests under our EUA authority as soon as we receive data regarding the accuracy and precision of the test from a diagnostic manufacturer to enable such authorization.

Vaccines and Therapies

There are no vaccines or treatments in advanced development for Zika virus at this time. Development programs for investigational vaccines and therapies are in the very early stages and FDA is actively engaged with NIH and BARDA to help accelerate development programs. FDA stands ready to work with medical product developers to provide technical support and clarify regulatory and data requirements necessary to move products forward in development as quickly as possible.

Vector Control

In the United States, mosquito control is typically achieved by a multi-faceted approach that includes a range of tools, including surveillance of mosquito activity, reduction in breeding sites, and the use of chemical and biological control methods. There has been public discussion of a new method to potentially help control mosquito populations through the use of a genetically engineered (GE) line of the mosquito *Aedes aegypti* (OX513A) developed by Oxitec, Ltd. The release of male Oxitec GE mosquitoes is intended to cause suppression of the mosquito population in a release area over time because the offspring resulting from the mating of male GE mosquitoes with wild type females do not develop to adulthood. The Brazilian National Biosafety Commission found the Oxitec GE mosquito safe for use in Brazil in 2014 after extensive field testing, and open field trials of these GE mosquitoes have also been conducted in the Cayman Islands, Panama, and Malaysia. Although multiple pilot studies have been authorized in Brazil, the country's national regulatory authority, the Agência Nacional de Vigilância Sanitária (ANVISA), is currently reviewing Oxitec's request to approve its product for commercial sale.

FDA is reviewing information in an Investigational New Animal Drug (INAD) file from Oxitec regarding the potential use of the company's GE mosquito with the intent of suppressing the population of *Aedes aegypti* mosquitoes at the release site(s). FDA ordinarily cannot acknowledge or discuss INAD files due to confidentiality concerns; however, FDA is able to do so in this case because Oxitec has publicly announced that they have opened an INAD file.

Oxitec is seeking to conduct a field trial to determine whether the release of its GE mosquito will suppress the local *Aedes aegypti* mosquito population in the release area at Key Haven, Florida. As with similar applications, FDA plans to release for public comment a draft environmental assessment submitted by Oxitec that assesses the potential environmental impacts of conducting a field trial in Key Haven with the GE mosquitoes. A field trial of the GE mosquito may only be authorized if—after a draft environmental assessment has been released for public comment and the Agency has had the opportunity to review those comments—FDA issues a final environmental assessment and Finding of No Significant Impact on the environment, or FDA prepares an Environmental Impact Statement. If such a trial is authorized and conducted, Oxitec could use the results from a field trial in an application for approval if the company decides to pursue one in the future.

Fraudulent Product Claims

Unfortunately, during emerging infectious disease outbreaks such as this, fraudulent products that claim to prevent, treat, or cure a disease rapidly appear on the market. FDA is actively monitoring for fraudulent products and false product claims related to Zika virus and will implement enforcement actions, as warranted, to protect the public health.

CONCLUSION

FDA is fully committed to remaining highly responsive and adaptive to the complex range of issues the Zika virus outbreak has presented and will continue to present. Developing the medical products necessary to help bring this outbreak under control is highly complex and will, unfortunately, take time. Close cooperation and collaboration within FDA, within the U.S. Government, with our international partners, and with product developers, is essential to help facilitate the development and availability of medical products to respond to Zika virus. There is an urgent need to accelerate medical product development programs, and FDA anticipates that—similar to the Ebola epidemic in West Africa—there will be a strong desire for access to investigational medical products as soon as they become available.

While the Ebola and Zika viruses have important differences, the response to the Ebola epidemic demonstrated that the typical, phased approach to medical product development can be accelerated and streamlined as appropriate, for the particular health circumstance; it may be appropriate in certain emergency circumstances to accept a greater than usual degree of uncertainty and risk in order to move rapidly to clinical trials, with the goal of getting safe, effective therapies or vaccine products to patients sooner. However, while greater risks may be acceptable during a public health emergency or when there is the potential for a public health emergency, the responsibility to protect patient welfare never ceases. The Ebola epidemic highlighted the importance of determining whether investigational products may help patients, do nothing, or cause unintentional harm. Thus, it is essential that clinical trials be designed to provide interpretable data— with respect to both safety and efficacy—that will enable the global community to learn whether investigational products are safe and effective for broader use.

FDA is fully prepared to leverage its authorities to the fullest extent practicable to help accelerate the development and availability of safe and effective products with the potential to help mitigate the Zika virus outbreak as quickly as the science will allow.

Thank you. I am happy to answer your questions.

In: The Zika Virus
Editor: Veronica Salazar

ISBN: 978-1-53610-124-9

Chapter 8

TESTIMONY OF DR. THOMAS R. FRIEDEN, DIRECTOR, CENTERS FOR DISEASE CONTROL AND PREVENTION. HEARING ON "EXAMINING THE U.S. PUBLIC HEALTH RESPONSE TO THE ZIKA VIRUS"*

INTRODUCTION

Good morning Chairman Murphy, Ranking Member DeGette, and members of the Subcommittee. Thank you for the opportunity to testify before you today on Centers for Disease Control and Prevention's (CDC's) efforts to prepare for and respond to the Zika virus outbreak, which threatens the United States and the rest of the Americas.

The Administration has requested approximately $1.9 billion in emergency funding to respond to the Zika virus outbreak, including $828 million for CDC, in support of both the domestic and international response, with particular attention to emergency assistance to the Commonwealth of Puerto Rico and other U.S. Territories and States with local transmission of Zika virus.

CDC is the nation's health protection agency, working 24-7 to save lives and protect people against unpredictable threats such as the Zika virus. Nature

* This is an edited, reformatted and augmented version of testimony presented March 2, 2016 before the House Committee on Energy and Commerce, Subcommittee on Oversight and Investigations.

is a formidable adversary, and Zika is our newest threat, particularly to pregnant women. CDC has some of the world's leading experts both in diseases spread by mosquitos and in birth defects. We must act swiftly to stop the spread of the Zika virus, both domestically and globally. While we are learning more about the Zika virus every day, there are many things we do not know yet about Zika. These include our understanding of the spectrum of effects of Zika infection during pregnancy, the risk the virus may play in microcephaly, Guillain-Barré syndrome and other possible complications, the duration of Zika infectivity in semen, and determining what other factors may play a part in the consequences associated with the virus. In addition to answering these questions, we are also working to accelerate optimal mosquito control strategies, improve testing and assure preparedness for rapid detection, control, and prevention within the United States and U.S. territories.

We are making advancements in these areas and will need the additional requested funding to do so. We are figuring out more about Zika literally every day, and will share information – and adjust our guidelines and recommendations – as we learn more. That is the nature of a scientific response to an emerging health threat. The doctors, scientists, entomologists, and others at CDC are working nonstop to protect Americans from this and other health threats. We have already made significant progress identifying the Zika virus in brain tissue of affected deceased infants, developing new diagnostic tests, issuing guidance, conducting epidemiological investigations along with affected countries, and improving monitoring and surveillance in the United States including in the Commonwealth of Puerto Rico and the other U.S. territories. Much of what we know about Zika and similar viruses today is based on the work that's been done by CDC scientists. But there are still many things we do not yet know. We will continue to use the best of modern science to protect the American people. I understand that Zika virus and the emergence of serious birth defects cause concern. We are committed to providing the American people with the most accurate and timely information about Zika virus and the current outbreak.

CDC is working in collaboration with other components of the Department of Health and Human Services (HHS), including the Office of the Assistant Secretary for Preparedness and Response (ASPR) and its Biomedical Advanced Research and Development Authority (BARDA), the National Institutes of Health, and the Food and Drug Administration (FDA). We are also working with partners across the U.S. Government to communicate with travelers and health care providers, update travel alerts and clinical guidance, and develop improved mosquito-control methods.

Zika and Its History

Zika is a flavivirus, which is closely related to dengue, yellow fever and West Nile viruses. Zika virus is primarily spread to people through the bite of infected *Aedes* species mosquitos, particularly *Aedes aegypti*. The *Aedes aegypti* mosquitos, which also transmit dengue and chikungunya viruses, are extremely difficult to control. They bite during the day, indoors and outdoors, and they preferentially feed on humans. And they need only the smallest bit of water to breed – just a bottle cap is enough. The mosquitos become infected when they bite a person with Zika virus. These infected mosquitos can then spread the virus to other people through bites. Case reports of other modes of transmission include spread through sexual transmission and blood transfusion. Of great concern, Zika virus infection in a pregnant woman has been linked to issues in fetal development, and the virus has been detected in association with fatal brain malformation in newborns as well as in miscarriages.

While its adverse effects were unforeseen, Zika is not a new virus. It was first recognized in 1947 and has caused occasional illness in Africa and Asia, but the first outbreak we know of occurred in 2007 in the small Pacific island of Yap. Last May, the first local transmission of Zika in the Americas was reported in Brazil, and by the end of 2015, Brazilian authorities estimated that the outbreak there involved perhaps a million suspected cases of Zika virus. In recent months, the virus has spread rapidly throughout Latin America and the Caribbean, as well as to parts of the Pacific. As of February 18, 2016, 32 countries and territories, including the Commonwealth of Puerto Rico, a United States Territory, the U.S. Virgin Islands, and American Samoa have reported local transmission of the Zika virus.

Symptoms and Adverse Outcomes

Many people exposed to Zika virus will have only mild symptoms - such as fever, rash, joint pain, and red eyes or conjunctivitis - that will last no more than a week. In past outbreaks, about four out of five people infected with Zika appear not to have had symptoms at all, although we do not know if that is the pattern in this outbreak.

Increasing evidence suggests that Zika virus infection may be associated with more serious health outcomes. In October 2015, Brazilian authorities recognized a concerning increase in microcephaly, which has occurred in close sequence to Brazil's outbreak of Zika virus. Microcephaly is a usually rare, serious condition where a baby's head is smaller than expected based on age and sex. Microcephaly is not a diagnosis in and of itself, but a sign that the brain did not develop as it should in the womb. Babies with microcephaly can have a range of problems, including seizures, developmental delay, feeding problems and hearing loss. In some cases these problems can be fatal.

Laboratory tests at CDC strongly suggest a link between Zika virus infection during pregnancy and microcephaly.

We do not fully understand the nature of this relationship, or if there are important cofactors. We also do not know what, if any, other outcomes might be associated with Zika infection during pregnancy among infants who do not have microcephaly. Microcephaly in infants can be devastating to the affected families, and this ongoing outbreak is concerning to everyone, especially for pregnant women, and their families who may travel to or live in the infected areas. The association between Zika virus and microcephaly is unexpected. A new infectious cause of fetal malformations has not been identified in decades. Zika virus spread in the Americas and its effect on pregnancy are developments that we are working with partners to better understand.

Our key priority at this point is to reduce the risk to pregnant women of Zika virus infection. Given the potential risks associated with maternal Zika infection, prevention is key for this response, with a parallel approach of acting based on what we know now and, at the same time, discovering more so that we can better prevent adverse health outcomes in the future. That's why, during the same week we identified Zika in brain tissue specimens from affected infants, we issued a warning to advise pregnant women not to travel to affected areas. That's why we are working intensively with the Commonwealth of Puerto Rico and other areas to get support to women who are or who may become pregnant and do what we can to reduce the threat of Zika there. And that's why we are engaging in studies with international partners so that we can more fully understand the magnitude of risk and the range of outcomes associated with Zika virus infection during pregnancy.

Health authorities in Brazil and elsewhere have also reported an increase in suspected cases of Guillain-Barré syndrome, a rare neurologic disorder in which a person's own immune system damages nerve cells, leading to nerve damage or paralysis that lasts for several weeks or several months. Most people fully recover, but it can take a few months or even years to do so. Some people with Guillain-Barré syndrome have permanent damage and, in rare cases, people have died. It is difficult to determine if any particular pathogen "caused" or "triggered" Guillain-Barré syndrome. Currently, we do not know if Zika virus infection causes Guillain-Barré syndrome.

However, the development of Guillain-Barré syndrome is a recognized after-effect of a variety of different infections. CDC is currently collaborating with public health officials in Brazil to investigate whether there is any causal link between Zika infection and Guillain-Barré syndrome.

Domestic Activities

While we are working to better understand these health outcomes, transmission, diagnostics, and mosquito control, CDC is moving quickly to respond. We have moved our Emergency Operations Center to the highest alert level for Zika virus to further enhance our response activities in areas with current local transmission and to accelerate preparedness efforts in anticipation of local transmission in the continental United States.

For the Commonwealth of Puerto Rico as well as the U.S. Virgin Islands and American Samoa, a surge in resources is urgently needed. The population of *Aedes aegypti* mosquitos is widespread on these islands, protective environmental factors such as window screens are not as prominent, and the density of people puts people there at high risk for transmission. All three areas have already reported local Zika transmission, with Puerto Rico alone reporting at least 30 cases. Furthermore, recent outbreaks of dengue and chikungunya suggest that Zika virus may spread extensively and rapidly in these areas. CDC has deployed staff to the U.S. Virgin Islands, American Samoa, and Puerto Rico to support response activities and provide technical assistance to health departments there. CDC and the CDC Foundation are also partnering to create Zika prevention kits.

Containing educational materials, and initial supplies of prevention tools such as insect repellant, the purpose of these kits is to help pregnant women in areas with local Zika transmission protect themselves and their pregnancies. Five thousand of these kits have been dispatched to the Commonwealth of Puerto Rico, the U.S. Virgin Islands, and American Samoa; and CDC plans to distribute more than 45,000 kits to these areas in the future.

While we have not yet seen transmission of the Zika virus by mosquitos within the continental United States, we expect many returning travelers will have Zika infection. As a potential benchmark, we received reports of 3,270 travelers from 49 states with laboratory confirmed cases of chikungunya infection in 2014 and 2015. There are about 40 million people travelling between the continental U.S. and Zika-affected areas each year. Therefore, all U.S. jurisdictions must be prepared to evaluate, test, and manage patients with potential Zika virus infection, particularly pregnant women. Furthermore, *Aedes aegypti* is found in many areas of the United States, raising the risk of local transmission. The most recent data available suggest that *Aedes aegypti* are found in 13 states and *Aedes albopictus* are found in 31 states and the District of Columbia. Recent chikungunya and dengue clusters in the United States suggest that Zika outbreaks in the U.S. mainland may be relatively small and localized due to protective factors like window screens and less dense living conditions; however, any local outbreaks will be of deep concern to the people living there, and we must be prepared for different scenarios including more extensive transmission risk.

CDC is working with health departments across the country to ensure coordination and to expand capacity for detecting and responding to Zika virus. Surveillance is essential to monitor and quickly identify areas with local transmission. We conduct multi-faceted surveillance for arboviruses, including Zika, through ArboNET, an integrated network which funds, through our Epidemiology and Laboratory Capacity cooperative agreements, staff in 49 states, the Commonwealth of Puerto Rico, and six large municipalities to conduct human case investigations, collect and test mosquitos, and perform laboratory analysis on arboviruses including Zika. Zika virus is now a nationally notifiable disease, meaning states report the virus to CDC, which will aid Zika surveillance efforts. CDC is also working with several states and the Commonwealth of Puerto Rico to determine a baseline prevalence of microcephaly so that any increase, should it occur, can be quickly and accurately identified.

With support from the President's emergency request, CDC will build on its current efforts to provide financial and technical resources to states and territories through its cooperative agreements to strengthen their capacity to prepare for and respond to emerging insect-borne threats such as Zika virus. These resources may be used to help health departments expand their capability to manage cases of local Zika virus transmission in their areas and to implement community education and prevention programs to reduce human mosquito contact and subsequently, the risk of Zika transmission. Resources will also be used to implement mosquito control strategies, including mosquito surveillance. Current mosquito surveillance capacity is uneven across the country, which makes our knowledge about the locations of the two mosquito vectors that transmit Zika virus potentially incomplete. To effectively track the spread of the outbreak, it is critical that states and territories receive specimens and test for Zika virus to diagnose and report travel-related and locally acquired cases of Zika. Under the emergency request CDC will expand its efforts to assist public-health labs nationwide to test for Zika and to provide the guidance on how to interpret test results. In addition, CDC is available to provide testing of any Zika samples upon request.

We are working to expand the number of health departments that have the ability to perform testing, but will need to increase the existing capacity to meet the projected demand for Zika testing. Given that, last year, it is estimated that approximately 500,000 travelers to areas of current Zika transmission were pregnant women and 36,000 pregnant women are currently living in the Commonwealth of Puerto Rico, the expansion of testing capacity in public health labs nationwide, included in the request, is urgently needed in order to ensure that every pregnant woman needing testing for Zika virus has access.

Recognizing the potential for Zika virus transmission through blood transfusions, CDC is collaborating with FDA to ensure the safety of the blood supply from Zika virus, particularly in regions experiencing local outbreaks.

CDC has sent experts to the Commonwealth of Puerto Rico to assess the steps needed to assure both that Puerto Rico's blood supply needs are met and that transfusion-transmitted Zika is prevented.

CDC experts are working intensively to learn more about the outbreak and provide people with the information they need to protect themselves. We will continue to issue travel alerts for the affected areas as confirmation of the virus is reported, and we'll keep the American people informed as the situation changes. We recognize people are eager for information, and our website has exceeded half a million views in recent days.

CDC has also provided guidance for doctors and other clinicians on evaluation, treatment and follow-up care of pregnant women and infants with possible exposure to Zika virus, partnering with organizations from around the health care community to help distribute this information as widely as possible. Our guidance will continue to be updated as our knowledge increases. We have recently updated our guidance to provide recommendations for the clinical care and management of pregnant women living in areas where Zika transmission is widespread, with special consideration to the ongoing risk of maternal Zika virus infection throughout pregnancy. These guidance documents were prepared in consultation with the American College of Obstetricians and Gynecologists, the Society for Maternal Fetal-Medicine, and the America Academy of Pediatrics.

CDC also wants to ensure that the general public knows what it can do to protect itself. Pregnant women should postpone travel to regions with ongoing Zika virus transmission. If they must travel, or if they live in affected areas, CDC recommends pregnant women talk to their doctors or other healthcare providers first and strictly follow steps to prevent mosquito bites. Reducing exposure to mosquitos is important for anyone traveling to or residing in areas where the virus is circulating. Wearing long sleeves, long pants, using EPA-registered repellents such as DEET and permethrin-treated clothing (both of which are safe to use in pregnancy), and using other protections such as air-conditioning will reduce exposure to mosquito bites. Given the potential for Zika virus to be spread through sex, pregnant women and their male partners living in or who have been to Zika-affected areas should abstain from sex or use condoms for the duration of pregnancy. This is a rapidly changing situation and our understanding of the risks concerning Zika virus infection is incomplete and evolving. As we get new information, we will update our advice.

Global Activities

On February 1, the World Health Organization (WHO) declared the recent cluster of microcephaly cases and other neurological disorders (such as Guillain-Barré syndrome) and their possible association with Zika virus, a public health emergency of international concern, a reflection of the seriousness of this unfolding health threat.

CDC is coordinating its response with the U.S. Agency for International Development, as well as the Pan American Health Organization (PAHO), the regional arm of the World Health Organization (WHO), and other parts of WHO, and is collaborating with many international partners to learn more about this outbreak. We are working with the Brazilian Ministry of Health on investigation and research partnerships. Specifically, one partnership involves studying the link between Zika virus infection and microcephaly, while another is examining the relationship between Zika virus and Guillain-Barré syndrome. Research teams from CDC are also in other countries, including Colombia, to explore collaborations that will shed light on the risk of microcephaly in relation to Zika virus infection during pregnancy.

In addition, CDC is offering support to all countries so that they can test samples from microcephaly cases for serologic evidence of Zika virus infection, and CDC is helping countries throughout the Americas establish incountry diagnostic capacity. To that end, we are currently, and in conjunction with PAHO, providing training to laboratorians in South and Central America on diagnostic tests, including two recent workshops in Brazil and Nicaragua.

CDC's Central American office has also facilitated the verification of Zika cases in several countries throughout Latin America, including Colombia, Venezuela, and Nicaragua. At the request of the Department of State's Bureau of Medical Services, staff from CDC's Global Disease Detection Center in Guatemala has been involved in communication efforts to ensure that new information regarding Zika virus and its possible link to birth defects is communicated to U.S. Mission Health Unit staff throughout the Americas.

The Global Health Security Agenda, with critical support from Congress, is collaborating with countries around the world so that we can find, stop, and prevent health threats when and where they first emerge. Zika has been present in Africa for decades, and it's possible that it could become linked to microcephaly there as well. The sooner we detect a problem, wherever it occurs, the more rapidly we can respond to it and prevent it from spreading. It is in all of our best interests to work with others to improve public health capacity around the world.

Improving the Tools and Information for Responding to Zika

We need a better understanding of the epidemiology of Zika and potential Zika-associated birth defects and other adverse health outcomes. We need better diagnostic methods that can quickly and clearly differentiate between similar viruses to detect evidence of past Zika infection. Testing for current Zika infection is only reliable in the first week of illness. A Reverse Transcription-Polymerase Chain Reaction (RT-PCR) test can provide a definitive diagnosis of Zika, but only if it is performed within about seven days of symptom onset. The tests we have available for Zika in persons who are no longer ill may have cross-reactivity with similar flaviviruses, particularly dengue, which can lead to false-positive or inconclusive results and confirmatory testing is required. Diagnosis is particularly challenging with Zika virus since most people will not experience symptoms. We also need to determine how long a man who has been infected with Zika may continue to be able to sexually transmit the virus to a partner, and we need better tools to screen the blood supply.

We also need to advance our ability to control the mosquito population. Existing methods for mosquito control all have shortcomings, especially in areas where the population of *Aedes* mosquitos is rampant. Furthermore, in some areas like the Commonwealth of Puerto Rico, mosquitos may have developed resistance to certain insecticides, which could reduce the range of substances that can be used to effectively decrease mosquito populations. We need to implement the best tools we have today, improve current vector control strategies, and identify better options. We also need better mosquito surveillance to determine the location of mosquitos and areas with mosquito resistance to insecticides, which would inform the implementation of new mosquito control techniques.

Finally, a vaccine is needed to protect people at risk of Zika virus infections, particularly preventing infection among women of childbearing age. At CDC, our scientists developed both a West Nile virus vaccine, which is currently in use for animal protection in the United States, and a dengue vaccine, which is currently in clinical trials. The President's request will increase Zika research, improve diagnostics and support advancements in vector control methods. Although availability of a licensed Zika vaccine is several years away, we do not know how long Zika will be a problem in the Americas nor whether the mosquito control efforts that must be implemented will yield durable results.

Conclusion

Microbes continue to be formidable adversaries. To protect Americans, the Zika emergency request invests in the laboratories, disease detectives, disease tracking systems, mosquito control, and investigations needed to continue to improve these essential tools.

The emergence and reemergence of health threats, including those spread by mosquitos and other vectors is not a unique event but something we expect to continue to see in the future. These outbreaks cannot be expected to occur in isolation of one another. The Commonwealth of Puerto Rico and Hawaii were already responding to outbreaks of dengue when Zika virus arose as an urgent health threat. We need to address the threat of mosquitoborne diseases systematically, rather than episodically. Thank you again for the opportunity to appear before you today. I appreciate your attention to this concerning outbreak and I look forward to answering your questions.

In: The Zika Virus
Editor: Veronica Salazar
ISBN: 978-1-53610-124-9

Chapter 9

TESTIMONY OF LAWRENCE O. GOSTIN, FACULTY DIRECTOR, O'NEILL INSTITUTE FOR NATIONAL AND GLOBAL HEALTH LAW. HEARING ON "EXAMINING THE U.S. PUBLIC HEALTH RESPONSE TO THE ZIKA VIRUS"*

EXECUTIVE SUMMARY

Following the 2014/15 Ebola epidemic, the World Health Organization (WHO) was widely criticized for its delay in taking decisive action and leadership to protect global health. Four international panels found that the WHO unreasonably delayed declaring Ebola a Public Health Emergency of International Concern (PHEIC) to call the international community to action in providing financial and technical resources for the global health humanitarian and security threat posed by Ebola.

The WHO appeared to learn from the Ebola experience. On February 1st, 2016, the Director-General, Margaret Chan, declared the clusters of microcephaly and neurological syndromes (including Guillain-Barré Syndrome) associated with Zika virus infection a PHEIC. The link between Zika virus infection and microcephaly and Guillain-Barré Syndrome is not yet proven, although suspected. Now the international community, including the

* This is an edited, reformatted and augmented version of testimony presented March 2, 2016 before the House Committee on Energy and Commerce, Subcommittee on Oversight and Investigations.

United States, must answer the call to action to ensure global health security and equity against the risk posed by the spread of Zika virus.

As the only developed country in the Americas with the *Aedes aegypti* mosquito, the United States bears not only a unique responsibility to take leadership for global health in our region, but also to protect the health of United States residents. The Zika virus is already embedded in Puerto Rico, and it is likely that by the summer time there will be local transmissions in Florida and the Gulf Coast. During the summer, the *Aedes aegpyti* mosquito has been observed as far north as Washington, DC,[1] and the *Aedes albopictus* mosquito, believed to also transmit the Zika virus, is found in at least 32 states in the United States.[2]

As Ebola demonstrated, the health security of the United States is intrinsically tied to ensuring strong health systems and a coordinated international response at the source of the outbreak. Furthermore, as a global leader for humanitarian action, the United States can provide the financial and technical resources necessary to address the Zika epidemic and protect global health. The work of the United States military and the Centers for Disease Control and Prevention (CDC) during Ebola proved to be transformative. The Ebola response was made possible by a special appropriation during the height of the epidemic, as requested by the President and allocated by Congress.

I fully support President Obama's FY2016 emergency supplemental appropriations request of $1.86 billion to respond to the Zika virus domestically and internationally. This must be in addition to existing global health security funding – including unobligated Ebola funding – to ensure flexibility to respond to the United States' needs as we progress towards Summer and the threat of domestic mosquito-borne transmission.

The Global Response to the Zika Epidemic

I have served on several high-level panels reviewing the WHO and international community's response to the 2014/15 Ebola outbreak including on the Harvard-London Hygiene and Tropical Medicine Independent Panel on the Global Response to Ebola and the National Academy of Sciences Global Health Risk Framework Commission. I also closely advised the United Nations Secretary General's High Level Panel on Global Health Crises. The WHO's failure to act quickly and decisively cost thousands of lives.[3] Of particular concern was the unreasonable delay in the convening of an Emergency Committee to advise to the WHO Director-General whether she should declare the Ebola outbreak a Public Health Emergency of International Concern (PHEIC) under the International Health Regulations (2005). The Director-General was also late in mobilizing the international financing needed to prevent international spread of the Ebola Virus.

Despite these critiques from numerous independent panels as well as implementing internal reforms, the WHO delayed taking its leadership role in response to Zika. The CDC and Pan American Health Organization had been working actively on Zika, in collaboration with the Brazilian government for months prior to the WHO's active engagement. Given the broad, global distribution of the mosquitoes that transmit Zika virus, and in the absence of rapid diagnostic tests or a vaccine, on January 27, 2016, my colleague Daniel Lucey MD and I called for the WHO to not repeat its mistakes from Ebola, and for the WHO Director-General to immediately convene an Emergency Committee on Zika virus to assess whether a PHEIC should be declared.[4] On January 28, 2016, the WHO announced that the Director-General would convene an Emergency Committee on Zika virus and the observed increase in neurological disorders and neonatal malformations.[5]

Following this announcement, the WHO acted decisively and with leadership. On February 1st, 2016, the WHO Director-General Margaret Chan declared the clusters of microcephaly and neurological syndromes including Guillain-Barré Syndrome a PHEIC.[6] This was the fourth PHEIC the WHO Director-General has declared since the International Health Regulations (2005) came into effect in June 2007, including for Influenza A (H1N1), polio, and Ebola.[7] While a causal relationship between these clusters and Zika virus is strongly suspected, it is not yet scientifically proven. Nonetheless, the declaration of a PHEIC reflects the pressing need for the coordination and resourcing of international efforts to address the Zika epidemic.

Accompanying the PHEIC, the Director-General issued temporary recommendations to countries on handling the Zika epidemic. While not legally binding, these temporary recommendations have significant normative weight and are issued in accordance with the legally binding International Health Regulations (2005). The recommendations advise countries to enhance surveillance for Zika virus infections, develop diagnostic tools, communicate the risks of Zika virus infection, aggressively implement vector control and personal protective measures, and ensure reproductive health services, including access to contraception (including emergency contraception) and maternal health care.

On February 16th, 2016, the WHO released its "Zika Strategic Response Framework and Joint Operations Plan".[8] This plan is a critically important step towards an effective response to the Zika epidemic. It is also a critical step towards re-establishing the WHO's global health credibility which was significantly undermined during Ebola. The plan estimates that $56 million will be required for the Zika epidemic response.[9] The majority of the funding is for risk communication and community engagement ($15.5 million) and care for those affected ($14.2 million). Given the vital role that the Pan American Health Organization (PAHO) plays in leading the global health response in the Americas, a large amount of the funding under this plan is appropriately earmarked for PAHO ($8.1m), including $1.5 million for surveillance and $2.25 million for vector control measures. Yet, despite this estimate of the needs, the WHO has yet to mobilize the required funding. That is still another reason to support the President's proposed emergency Zika appropriation.

While the WHO's commitment to raising funds needed for the global response is positive, the estimate of $56 million is far less than that required to stem the Zika epidemic in the Americas.

Major global funding is needed for aggressive mosquito control, surveillance, and research. Under the WHO's plan, the overall estimates for these strategies are $6.4 million, $7.1 million, and $6.4 million respectively. The Zika epidemic is likely to spread to many regions of the world beyond the Americas, which will require considerably greater funding. During the Ebola outbreak, the WHO released financing estimates that were far from adequate, requiring repeated increased cost estimates as the Ebola outbreak unfolded. Overly conservative estimates for the Zika epidemic response could result in the same mistakes seen with Ebola: waiting until a global emergency to mobilize funding will always be too little, too late. The WHO must have access to a much larger emergency contingency fund. The trigger for use of this fund would be the declaration of a PHEIC.

The Global Health Risk Framework Commission (on which I served) estimated that it would cost $4.5 billion annually to truly reduce pandemic risks.[10] While this may appear to be a great deal of money, it pales in comparison to the global costs of a pandemic. Previous epidemics have demonstrated that states with major emerging disease outbreaks experience a loss in GDP of up to 10%. Brazil is may experience a similar economic impact with the upcoming 2016 Olympics and Paralympics in Rio de Janeiro. The World Bank has advised that while initial estimates of the short-term economic impact of the Zika epidemic in the Latin American and Caribbean region is 0.06% of GDP, such estimates are based on a swift, coordinated international response. The World Bank however advises such estimates could be significantly larger given the high dependence on tourism in the region if the virus is not promptly contained, if a causal association between Zika virus and clusters of microcephaly and Guillain-Barré Syndrome is established, and if public perceptions of the risks of Zika change.[11]

In addition to the financial and technical resources dedicated by the WHO, it is vital that the WHO use its normative power for global public health. The WHO was criticized for prioritizing politics over public health during the Ebola outbreak. It must not let the same happen with the Zika epidemic, which disproportionately burdens women and the poor. In particular, calls by governments that women avoid getting pregnant – including Brazil, Honduras, Colombia, and El Salvador – without providing access to reproductive health and rights services, are not only ineffective but also unfair to women. In Central and South America, there is limited access to contraception – especially in poorer communities most at risk of Zika virus infection and cultural and religious norms still restrict the availability and use of contraception. Nearly all countries in the Americas currently experiencing or likely to experience Zika virus cases have laws preventing women from accessing reproductive health services such as emergency contraception and abortion. In addition, the quality, accessibility, and affordability of maternal health care systems vary greatly between and within countries, with varying degrees of post-partum support services for children born with disabilities. Pope Francis recently supported the expanded use of contraception in the face of potential harms to pregnant women from Zika virus infection.

Both the WHO and the international community's response to the Zika epidemic must therefore be comprehensive, equitable, and substantial.

The United States' Domestic Response

The Risk to the United States

The mosquitoes that transmits Zika virus – *Aedes aegypti* (primary vector) and *Aedes albopictus* – is present in the United States and its territories.[12] One recent study modeling possible Zika spread estimates that more than 60% of the US population lives in areas conducive to seasonal Zika virus transmission,[13] with Florida and Southern Texas at risk of year round transmission. The United States is the only high-income country in the Americas at risk of local Zika virus transmission.[14] As a result, the United States domestic response must cover not only returning travelers who may be infected and local transmission of Zika virus from sexual intercourse, but also local transmission of Zika virus due to infected mosquitoes.

On February 3, 2016, the CDC's Emergency Operations Center (EOC) moved into Level 1 activation to accelerate domestic preparedness in anticipation of local Zika virus transmission by mosquitoes in the Continental United States.[15] Level 1 is the EOC's highest level of activation, reserved for critical emergencies. This involves assignment of the largest number of staff possible to work on a response. There have been three previous EOC Level 1 responses: the Ebola outbreak, Influenza A H1N1 outbreak, and Hurricane Katrina. Moving to this level of response demonstrates the importance of preparedness for the Zika outbreak and supports the development of laboratory tests to diagnose Zika infection, studies to investigate the association between Zika virus and microcephaly and Guillain Barré syndrome, surveillance in the United States and its territories, and support of efforts in Puerto Rico, Brazil, and Colombia.

Travel Advisories and Restrictions

The CDC has issued Level 2 (Alert: Practice enhanced precautions) general travel advisories for Zika virus in areas with local transmission such as South America, Central America, Mexico, Pacific Islands, and the Caribbean.[16] In addition, these Level 2 travel alerts include specific interim recommendations that pregnant women should consider postponing any travel to affected regions. The CDC has not issued any Level 3 (Warning: Avoid all non-essential travel) travel advisories which would indicate danger to all travelers, not only travelers with specific risk factors, such as pregnancy.

Travel advisories are important for ensuring that individuals traveling to Zika virus affected countries have adequate information to decide whether to travel and what steps they should take to prevent Zika infection while in affected countries and upon return to the United States. Given the typically mild symptoms of Zika virus in healthy, non-pregnant individuals, Level 3 travel warnings are not currently warranted. Priority should be given to continuing the current information sharing process to travelers on how to prevent mosquito bites, such as through the use of EPA-registered insect repellents, the covering of exposed skin, and sleeping in screened or air-conditioned rooms, and how to prevent possible infection of pregnant partners through sexual intercourse upon return to the United States.

Notably, the CDC and WHO have different advice to pregnant women traveling to Zika affected countries. The CDC's response is more proactive and one that I strongly support.

Travel bans are not warranted, including specific bans for pregnant women at any stage during pregnancy, which would be unnecessarily discriminatory and intrusive. It is important that the CDC is supported in its efforts to set, continually review, and update travel advisories as the public health circumstances change. The WHO Director-General's temporary recommendations issued with the PHEIC declaration specifically advise against travel or trade restrictions to prevent the spread of Zika virus.

Quarantine and Isolation

As a vector-borne disease, Zika virus is predominately transmitted through the bite of an infected female *Aedes aegypti* mosquito. In addition, most individuals infected with Zika do not show symptoms, making identification of individuals for isolation very difficult. The quarantine of persons suspected of Zika virus infection, or isolation of persons with Zika virus infection, is therefore not an effective or warranted public health strategy. While there is evidence of sexual transmission of Zika, this mode of transmission would also not be favorable to quarantine or isolation measures. As a result, the use of quarantine and isolation in response to Zika is unfounded and therefore likely to be considered an unreasonable and unnecessary burden on individuals' liberty that does not have justification in public health. In the temporary recommendations issued with the PHEIC, the WHO Director-General has advised against restrictions such as quarantine and isolation to prevent the spread of Zika virus. Control efforts should therefore focus on mosquito eradication, education about the measures individuals can take to protect themselves from infection, and funding the development of a vaccine.

Vector-Control

It is a public health priority that the main source of Zika virus transmission – principally the *Aedes aegypti* mosquito – is subject to aggressive vector control efforts. In addition to the physical removal of stagnant water that serves as mosquito breeding sites, the use of safe, effective and approved insecticides is necessary. However, insecticide resistance is growing and the WHO is investigating novel mosquito control strategies, such as the use of transgenic mosquitoes to reduce local mosquito populations.[17]

Given the public health impact of the *Aedes aegypti* mosquito, I support the watchful use of transgenic mosquitoes through a carefully controlled release that monitors the effects on mosquito-borne diseases – including Zika, Dengue, and Chikungunya viruses –as well as any potential ecological harms.

Trials so far have demonstrated promising results in drastically reducing local mosquito populations that carry these diseases without causing other harms. There is also little evidence that transgenic mosquitos cause ecological harms, although potential harms need to be closely watched and controlled for.

Ensuring Equity and Justice in the United States' Health System

Given the disproportionate impact of the Zika epidemic on women, it is essential that women in the United States are guaranteed access to reproductive health services including contraception, emergency contraception, and maternal healthcare. Where it is lawful, pregnant women with a fetus known to have microcephaly should be offered the opportunity to abort the pregnancy.

Given the possible sexual transmission of Zika virus, condoms must be readily available and accessible to both men and women in the United States. In addition, concerns about fetal health cannot stop at birth: infant, child, and ongoing healthcare and support services for babies that may be affected by the Zika epidemic must be guaranteed. The Federal and States Governments must act to protect the health of US citizens, and meet their obligations under international health and human rights law.

Ensuring Financing for Response Efforts: President Obama's Supplementary Funding Request

On February 22nd, 2016, President Obama submitted a FY 2016 emergency supplemental appropriations request of $1.86 billion to respond to the Zika virus domestically and internationally.[18] The vast majority of this funding request is directed to the Department of Health and Human Services – $1.48 billion, $828 million of which is directed to the CDC. This funding includes $200 million to support the development of a vaccine through to commercialization with the National Institutes of Health and product development and approvals with the Food and Drug Administration. The United States Agency for International Development (USAID) would receive $335 million for targeted activities in South America, Central America, and the Caribbean. Finally, $41 million would be directed to the Department of State to *inter alia* support WHO, PAHO, and UNICEF public health efforts to address the Zika virus in affected countries while ensuring United States health security. Given the unknowns of the health humanitarian and security risk posed by Zika – especially as we transition to Summer – this funding is essential to ensure flexibility to respond to the United States' needs, including beyond that already identified.

Use of Unobligated Ebola Emergency Funds for the Zika Response

While it may seem appealing to use money previously allocated to the Ebola outbreak, such a move risks making the United States more vulnerable to infectious disease threats. Failure to develop and secure health systems in developing countries – whether in West Africa or South America – risks the health, safety, and security of the United States.

Ensuring global public health is not simply about a rapid response, but a sustainable response. Reallocating money from one infectious disease outbreak to another undermines efforts to ensure surveillance, detection, response, and control strategies are not only operational but sustainable in countries at risk of outbreaks. While infectious diseases may appear quickly, the efforts to prevent and respond to them cannot simply be about extinguishing fires, but rather preventing them from flaring up and occurring again in the future. Given the economic impact of outbreaks, sustainable and systems-wide technical and financial resources are essential to ensure taxpayers' money is utilized effectively. Furthermore, reallocating money from one infectious disease outbreak to another infectious disease outbreak risks the long-term health, safety, and security of the United States. It is important to note that the USAID funding request in President Obama's FY2016 emergency supplemental appropriations request would provide flexibility in the use of remaining USAID Ebola funds.[19]

Global health security and public health is grossly underfunded. I strongly advise that the Zika supplementary budget request is funded independently from unobligated past emergency funds.

End Notes

[1] Andrew Lima et al., "Evidence for an Overwintering Population of Aedes Aegypti in Capitol Hill Neighborhood, Washington, DC," The American Journal of Tropical Medicine and Hygiene 94, no. 1 (January 6, 2016): 231–35, doi:10.4269/ajtmh.15-0351.

[2] Anthony S. Fauci and David M. Morens, "Zika Virus in the Americas — Yet Another Arbovirus Threat," New England Journal of Medicine 374, no. 7 (February 18, 2016): 601–4, doi:10.1056/NEJMp1600297.

[3] Suerie Moon et al., "Will Ebola Change the Game? Ten Essential Reforms before the next Pandemic. The Report of the Harvard-LSHTM Independent Panel on the Global Response to Ebola," The Lancet 386, no. 10009 (November 2015): 2204–21, doi:10.1016/S0140-6736(15)00946-0.

[4] Lucey DR and Gostin LO, "The Emerging Zika Pandemic: Enhancing Preparedness," JAMA, January 27, 2016, doi:10.1001/jama.2016.0904.

[5] WHO, "Emergency Committee on Zika virus and observed increase in neurological disorders and neonatal malformation" (January 28, 2016), available at: http://www.who.int/mediacentre/news/statements/2016/emergencycommittee-zika/en/ (accessed: February 24, 2016).

[6] WHO, "WHO statement on the first meeting of the International Health Regulations (2005) (IHR 2005) Emergency Committee on Zika virus and observed increase in neurological disorders and neonatal malformations" (February 1, 2016) available at: http://www.who.int/mediacentre/news/statements/2016/1st-emergency-committeezika/en/ (accessed: February 24, 2016).

[7] The previous declarations were the Influenza A H1N1 pandemic in April 2009, Polio resurgence in May 2014, and the Ebola outbreak in West Africa in August 2014.

[8] WHO, "Zika Strategic Response Framework and Joint Operations Plan" (February 16, 2016), available at: http://who.int/emergencies/zika-virus/strategic-response-framework.pdf?ua=1

[9] Specifically, $55,743,651: See, WHO, "Zika Strategic Response Framework and Joint Operations Plan" (February 16, 2016), available at: http://who.int/emergencies/zika-virus/strategic-response-framework.pdf?ua=1, Part III: Annexes.

[10] Commission on a Global Health Risk Framework for the Future and National Academy of Medicine, Secretariat, The Neglected Dimension of Global Security: A Framework to Counter Infectious Disease Crises (Washington, D.C.: National Academies Press, 2016), http://www.nap.edu/catalog/21891.

[11] World Bank, "The short-term economic costs of Zika in Latin America and the Caribbean (LCR)", (February 18, 2016), available at: http://pubdocs.worldbank.org/pubdocs/publicdoc/2016/2/410321455758564708/The-short-termeconomic-costs-of-Zika-in-LCR-final-doc-autores-feb-18.pdf (accessed February 24, 2016).

[12] CDC, "Surveillance and Control of Aedes aegypti and Aedes albopictus in the United States", available at: http://www.cdc.gov/chikungunya/resources/vector-control.html (accessed: February 24, 2016).

[13] Moritz U. G. Kraemer et al., "The Global Distribution of the Arbovirus Vectors Aedes Aegypti and Ae. Albopictus," eLife 4 (2015): e08347, doi:10.7554/eLife.08347 as cited in: Isaac I Bogoch et al., "Anticipating the International Spread of Zika Virus from Brazil," The Lancet 387, no. 10016 (January 29, 2016): 335–36, doi:10.1016/S0140-6736(16) 00080-5.

[14] Canada does not have the Aedes aegypti mosquito. Developmental status taken from: UN, World Economic Situation and Prospects Report 2014, available at: http://www.un.org/en/development/desa/policy/wesp/wesp_current/2014wesp_country_classification.pdf (accessed February 24, 2016)

[15] CDC, "CDC Emergency Operations Center moves to highest level of activation for Zika response" (February 3, 2016), available at: http://www.cdc.gov/media/releases/2016/s0208-zika-eoca-activation.html (accessed February 24, 2016)

[16] CDC, "Zika Travel Information", available at: http://wwwnc.cdc.gov/travel/page/zika-information (accessed February 23, 2016).

[17] WHO, "Mosquito control: can it stop Zika at source?" (February 17, 2016) available at: http://www.who.int/emergencies/zika-virus/articles/mosquito-control/en/(accessed: February 24, 2016).

[18] Letter from President Barack Obama to Speaker Paul Ryan (February 22, 2016), available at: https://www.whitehouse.gov/sites/default/files/omb/assets/budget_amendments/emergency_supplemental_2-22- 16_zika.pdf (accessed February 24, 2016)

[19] White House, Office of the Press Secretary, "Fact Sheet: Preparing for and Responding to the Zika Virus at Home and Abroad" (February 8, 2016), available at: https://www.whitehouse.

gov/the-press-office/2016/02/08/fact-sheetpreparing-and-responding-zika-virus-home-and-abroad (accessed February 24, 2016).

In: The Zika Virus
Editor: Veronica Salazar
ISBN: 978-1-53610-124-9

Chapter 10

TESTIMONY OF DR. KACEY C. ERNST, ASSOCIATE PROFESSOR, UNIVERSITY OF ARIZONA. HEARING ON "SCIENCE OF ZIKA: THE DNA OF AN EPIDEMIC"*

INTRODUCTION

Chairman Smith and Ranking Member Johnson, my name is Dr. Kacey C. Ernst and it is an honor to be providing this testimony. I am testifying about the current state of knowledge of the mosquito *Aedes aegypti* in the United States as it relates to Zika virus transmission and the current limitations and potential for forecasting Zika transmission in the United States.

I am an Associate Professor in the Department of Epidemiology and College of Public Health at the University of Arizona. I also hold positions in the Colleges of Geography and Animal and Comparative Biomedical Sciences, as well as in three interdisciplinary programs: Entomology and Insect Science; Global Change; and, Arid Lands. My area of specialization is the intersection of environment, humans, and mosquito vectors of disease. As an epidemiologist, my role is to work with a highly interdisciplinary team of scientists to integrate information from climatology, entomology, medical anthropology, and ecology to develop an understanding of the emergence of infectious diseases. I have conducted work on the *Aedes aegypti* mosquito since arriving at the University of Arizona in 2008.

---

* This is an edited, reformatted and augmented version of testimony presented May 25, 2016 before the House Committee on Science, Space, and Technology.

Prior to joining the faculty at the University of Arizona, I held positions in public health as an epidemiologist focused on pandemic threat reduction and bioterrorism preparedness. I graduated from the University of Michigan in 2006 with my doctorate in Epidemiology.

State of Knowledge of the *Aedes Aegypti* Mosquito as It Relates to Zika Virus Transmission

For mosquito-borne transmission of Zika virus, the following conditions must be met: there must be a susceptible human population, presence of the Zika virus, presence of a mosquito that is capable of transmitting the virus, and environmental conditions that allow the interaction of all three. Transmission potential is higher when: 1) there are higher densities of a mosquito that is biologically capable of transmission-- in this case *Ae. aegypti* and to a lesser extent *Ae. albopictus*, 2) these mosquitoes survive long enough to transmit the virus, and 3) the mosquitoes are feeding on humans at higher rates [1]. While seemingly simple, *understanding these components in relation to a specific virus can take decades of research*, as they are very context specific.

There is a long history of *Aedes aegypti* research that focuses on understanding these components, its biology, and interactions with arboviruses. Arboviruses are diseases carried or vectored by insects or related animals like ticks. Much of the previous work has focused on important viruses transmitted by *Ae. aegypti*, including dengue, yellow fever, and chikungunya, as Zika virus has only recently been perceived as a significant human threat. Much of what we currently assume about Zika virus is based upon the relationship between dengue viruses and *Ae. aegypti,* given they are closely related viruses that have been widely transmitted for hundreds of years [2]. However, the amount of knowledge specific to Zika virus is growing exponentially as the global health crisis has unfolded.

Aedes aegypti is found in many urban areas throughout the Southern United States and its potential range extends into the eastern seaboard (see Figure 1). Yellow fever outbreaks that occurred in the 1640's from southern Florida to New York City indicate that *Ae. aegypti* populations lived in these areas centuries ago. The decline in these diseases is attributed to improved living conditions, including piped water and sewers, and reduction of the introduction of *Ae. aegypti* through shipping due to changes in shipping routes [3]. *We do not really know the complete distribution of Ae. aegypti in the*

United States because surveillance for the mosquito is not consistent across jurisdictions, and many do not have the resources to carry out the mosquito surveillance needed to determine if Aedes species are established in their jurisdiction.

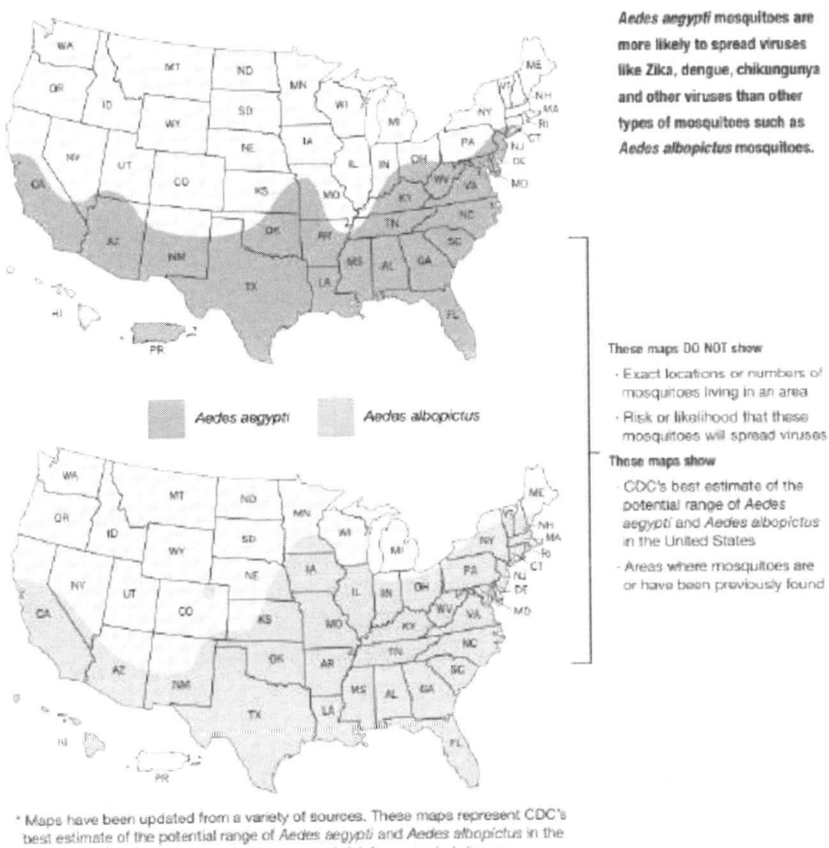

Source: CDC.gov.

Figure 1. Estimated range of *Ae. aegypti* and *Ae. albopictus* in the United States in 2016.

Ae. aegypti is a highly invasive species. It is originally from sub-Saharan Africa but has now spread throughout the warmer regions of the world, often hitchhiking in human-made containers like old tires. *This mosquito exploits the ways we have changed our environment*, including our use of many disposable containers, increasing movement of people and goods, and migration of much of the population to dense urban centers [4]. *Ae. aegypti* strongly prefers to feed on human blood and lives in and around our homes. As will likely be mentioned by others testifying here today, these mosquitoes are extremely difficult to control. The female mosquito lays her eggs above the water line on the side of the container, then the eggs are submersed in water when it rains or the container is filled by people. They can exploit extremely small pools of water and can mature from egg to adult within 7-10 days in only an inch of water [5]. *A single back yard may have dozens of containers such as buckets, birdbaths and old tires, which when filled with water by rain or human activities, like watering plants or storing drinking water, provide habitat for the mosquito.* Adult mosquitoes will also readily enter homes and can lay their eggs in vases, fish tanks and other sources of water. This tight link with humans and the built environment make them very efficient vectors of viruses such as Zika, dengue, chikungunya, and yellow fever. They are also known to carry other viruses, such as Mayaro virus, which are currently considered of minor importance, just as Zika virus was as recently as two years ago.

The transmission cycle of Zika virus between humans and *Aedes aegypti* mosquitoes is similar to the other viruses transmitted by mosquitoes. Most commonly, *transmission of Zika virus occurs when an infected person is bitten by a female mosquito of a particular species.* This includes female *Aedes aegypti* and a related species, *Aedes albopictus* (the Asian tiger mosquito, common in the Eastern United States). Once the mosquito has ingested the infected blood, it takes a specific period of time for the virus to move through the mosquito, replicate itself, and migrate back to the salivary glands. Only then can the transmission cycle be completed during the next blood meal taken by the *Ae. aegypti* female mosquito. As the mosquito feeds on the human blood, she injects virus-infected saliva into the person. The time period between when the mosquito first picks up the virus and when she can pass it to another person is known as the extrinsic incubation period, or *EIP*, and once the EIP is completed, the mosquito can transmit the virus for the rest of her life, which ranges between 2-4 weeks.

The shorter the EIP is, the faster the virus can be transmitted through a population. This is for two reasons: first, it reduces the time between potential infectious bites, and secondly, it increases the chances that a mosquito will be able to survive long enough to transmit the virus. The EIP is regulated by both the amount of virus ingested and temperature conditions. The warmer it is, the more rapid the EIP. *At this time, the EIP is not well-established for Zika viruses.* For dengue viruses, it is 7-10 days. The research group of which I am a member, led by Dr. Michael Riehle, has successfully competed for a grant from the National Science Foundation (NSF) to determine the EIP under different temperature conditions to determine how rapid the EIP is for Zika virus within the mosquito.

In addition to the primary Zika virus transmission cycle between mosquitoes and humans, there is a secondary transmission route in mosquitoes that may be possible, known as *vertical transmission.* Vertical transmission occurs when an infected female mosquito passes the virus directly to her offspring. In this scenario, the female mosquito will emerge from the egg already infected with the virus. When the female mosquito takes the first blood meal after emergence, it could already be capable of transmitting the virus to a human. There is some limited evidence that vertical transmission may occur, including a male mosquito that tested positive for Zika virus (only females feed on blood), and rapid seasonal onset of Zika virus transmission which might suggest the virus was harbored in eggs from the previous season [6, 7].

If vertical transmission of Zika virus proves to be common in Ae. aegypti populations, this could have implications for the speed of the spread and would mean that Zika virus might not have to be introduced each mosquito season. Eggs that were laid during one mosquito season could harbor the virus and emerge as adults the following mosquito season ready to transmit the virus. The role that vertical transmission may play in the Zika pandemic is currently unclear. For dengue viruses, vertical transmission is considered relatively uncommon and not a significant contributor to transmission [8]. As part of the experiments proposed for EIP, Dr. Riehle's group will also be conducting laboratory experiments to determine the frequency that it occurs in a laboratory setting. This would need to be coupled with evidence from the field, which has been proposed by colleague Dr. Kathleen Walker in a National Institutes of Health (NIH) grant application currently under review, to determine whether vertical transmission happens at a significant level during Zika outbreaks.

Key gaps in knowledge about the interactions between the Zika virus and the vector *Ae. aegypti* raise the following questions:

- Is being infected with the Zika virus detrimental to the *Ae. aegypti* mosquito (if the Zika virus shortens the *Ae. aegypti* lifespan, this would reduce transmission potential)?
- Are the sub-species of *Ae. aegypti* found through the United States equally competent vectors of the Zika virus compared to those sub-species attributed to the current pandemic in Latin America?

- What are the minimum infectious doses of virus required for the *Ae. aegypti* mosquito to become infected and subsequently infectious to humans?
- Does co-infection with other pathogens increase or decrease the transmission potential of Zika virus?
- Is the probability of transmission to and from the mosquito influenced by temperature?
- Can people who do not show symptoms of Zika virus transmit the virus to the *Ae. aegypti* and *Ae. albopictus*?

These are fundamental questions that need to be answered before accurate models of disease risk can be constructed. Recent evidence has been generated in laboratory studies from Dr. Thais Chouin-Carneiro of Instituto Oswaldo Cruz, a research center in Brazil, and Dr. Anubis Vega-Rua from Institut Pasteur that indicate *Ae. aegypti* and *Ae. albopictus* mosquitoes from the Americas have low vector competence, or the biological ability of the mosquito to transmit the virus, and attribute the large-scale outbreak to high mosquito numbers and human populations that are completely susceptible [9]. Additional research is needed to confirm these findings and several research groups in the United States have recently received pilot funding from the NSF to address some of these questions, including the research group of which I am a member, Dr. Courtney Murdock's group at the University of Georgia and Dr. Jefferson Vaughan at the University of North Dakota.

Other Mosquitoes of Potential Importance for Zika Transmission

It is broadly agreed that Ae. aegypti is the most important vector for transmission of Zika virus [10]. However, *Ae. albopictus mosquitoes, which have a broader potential range across the United States because they can survive cold weather better than Ae. aegypti, are also a capable of transmitting Zika virus*. Recently Zika virus was found in Ae. albopictus in Mexico, so they may also be transmitting Zika virus [11] Ae. albopictus has spread over the past decade in the continental United States and its potential range reaches further northward than Ae. aegypti. Ae. albopictus are more general feeders than Ae. Aegypti, which highly prefer human blood meals. Yet, evidence in the Northeastern United States demonstrates that Ae. albopictus also take a substantial proportion of blood meals from humans [12].

When a vector feeds primarily on humans, this increases the likelihood of transmission of viruses infecting human populations. These findings highlight *the importance of enhancing monitoring and surveillance for both Ae. albopictus and Ae. aegypti.*

The Potential and Limitations of Forecasting Zika Virus Transmission Risk Across the United States

Transmission of arboviruses, such as Zika virus, is driven by multiple factors, including environmental suitability to support the vector and the virus, interactions among vectors and the human population, and introduction of the virus into susceptible vector and human populations. Accurate predictions of transmission risk rely on the integration of information from all of these complex processes.

As part of our work to understand the risk of Zika virus infection in the continental United States, our team of scientists, led by Dr. Andrew Monaghan at the National Center for Atmospheric Research (NCAR), modeled seasonal dynamics of the *Ae. aegypti* mosquito vector across 50 cities in the Southern United States [13]. The intention of this map was *to provide information for decision-makers, such as yourselves, on the time periods and geographic areas with higher potential risk for local Zika virus transmission*. These maps depict the areas that are considered *climatically suitable* for the seasonal establishment of *Ae. aegypti* populations, but some of the 50 cities included in this analysis do not have confirmed *Ae. aegypti* populations. Because there is no defined threshold for "number of *Ae. aegypti* needed for transmission," we made a relative comparison between the predicted level of *Ae. aegypti* in 49 cities to the predicted levels in Miami, FL, one of the most climatically suitable areas in the continental United States that has a history of local transmission of other *Ae. aegypti*-borne viruses (dengue and chikungunya). It should further be noted that the standard of Miami, FL, is not the same as a standard from other areas where Zika virus is currently circulating as the southern United States lies at the cool margins of the range of *Ae. aegypti* [3].

The models used to predict the relative periods of climatic suitability for Ae. aegypti populations rely on our best, but incomplete, understanding of the processes by which temperature, precipitation, and humidity drive the dynamics of *Ae. aegypti*. Generally, more rainfall means more mosquitoes when man-made containers are present to hold water and provide habitat for the mosquito larvae. Temperature is important because warmer temperatures increase mosquito survival, and the development time between egg and adult mosquitoes becomes shorter, enabling populations to grow quickly. In addition, warmer temperatures shorten the time between blood meals taken by the *Ae. aegypti*. When warm temperature thresholds are exceeded, however, which occurs in some cities in the Southwestern United States during mid-summer, *Ae. aegypti* may not survive as well [14]. These relationships are complex and not all of them are well-defined, making modeling of *Ae. aegypti* populations somewhat uncertain.

To reduce this uncertainty our group used two models that have been validated previously: Skeeterbuster, developed from the original container-inhabiting mosquito simulation model (CiMSiM) by Dr. Dana Focks at University of Florida and refined by Dr. Fred Gould, Dr. Alun Lloyd and others at North Carolina State University [15], and the Dynamic Mosquito Simulation Model (DyMSiM), developed by Dr. Cory Morin, former University of Arizona graduate student and current postdoc at NASA Marshall Space Flight Center [16]. This combined modeling approach is useful when relationships are not well-defined. Our ensemble model appeared to capture the seasonality of the *Ae. aegypti* fairly well in the two locations that had data available, but there are discrepancies between the modeled and field data emphasizing the need to combine both modeling approaches and field collections. *These models will improve when ground-based surveillance for Aedes mosquitoes in the United States is more widespread and done at regular intervals.*

Climatic suitability and Ae. aegypti populations are critical factors but they alone cannot predict where transmission will occur. To further define high risk areas, we examined the geographic distribution of several important determining human factors: the number of returned travelers from countries with current Zika transmission (as of February 2016); counties in the continental United States that have reported locally-acquired cases of other *Aedes*-transmitted viruses – dengue and chikungunya – since 2010; and, the proportion of individuals who live in poverty by county in the continental United States during the summer months. The travel information is a rough indicator of the potential for viral introduction from countries where Zika virus is currently being transmitted. *An urban area which receives a large number of travelers from areas with transmission are more likely to have travel-related cases.* Given there are currently no locally-acquired cases, the only way that local transmission can occur is if a returned traveler brings the virus into the population. Counties where recent locally-acquired cases of dengue and chikungunya have occurred provide clues to where we might expect to see the emergence of Zika in the contiguous United States; these counties encompass areas of Southern Florida and Southern Texas.

Poverty is related to risk of transmission. Impoverished communities are at higher risk of Zika transmission for several reasons. Low-income neighborhoods with poorer infrastructure and sanitation tend to have more garbage (discarded containers that can serve as habitat for the immature mosquitoes), abandoned lots, and poorly maintained public areas in which water-holding containers that provide mosquitoes with larval habitat accumulate. In addition, homes may lack window or door screens that are intact, allowing the *Ae. aegypti* mosquito to more readily enter the house. And finally, lower income households often lack air conditioning or have lower quality air conditioning options compared to central air conditioning. Without central air conditioning during the summer months, individuals are more likely to keep doors and windows open, also increasing their contact with mosquitoes. Work by Dr. Mary Hayden (NCAR) also demonstrates that, even if individuals reside in homes with central air-conditioning, they may lack the funds to run or maintain the unit. This is *particularly problematic in the U.S.-Mexico border region, where the environment is highly suitable for the mosquito and poverty and crowded conditions are common*, and

there is high mobility between the U.S. and Mexico where Zika transmission is already occurring. The history of dengue outbreaks in this region supports the idea that this is a likely area for transmission of Zika to occur.

Another aspect of risk of particular interest to our research group is the *current infrastructure to support Ae. aegypti and Ae. albopictus surveillance and control.* We conducted an in depth search of publically available information to glean information on surveillance and control activities in the 50 cities included in our study. What we found, or rather did not find, was disheartening. *Very few jurisdictions actually publically reported any information on Aedes species* in the area and it was unclear which jurisdictions actually conducted surveillance for these mosquitoes at all. Given the extremely tight links between humans and the vector, *it is not only critically important to improve surveillance programs for these mosquitoes, but also to ensure that these surveillance data are transparent and available for communities.* To reduce *Aedes* species individuals, households, communities, and government administration must all be involved. People are better able to mobilize and control mosquito populations when they have up-to-date and specific information about the locations of mosquito populations.

It was not our intention to forecast where Zika virus would be present during 2016 in this study and, in fact, our simulations were based on average climatological conditions over the past decade, rather than meteorological forecasts for the upcoming summer season. Our intention was to quickly identify times of year and locations at higher risk with the best data available. *Actual forecasting of transmission is theoretically possible but requires significant improvement in our knowledge across several broad areas.* These include improved field data (including mosquito and disease surveillance) for validation, a better understanding of the interactions among humans and *Ae. aegypti*, more accurate seasonal weather forecasts, and data on the potential distribution and impact of vector control resources and activities on mosquito abundance.

We need better field data for validation of predictions. As noted above, surveillance for *Aedes aegypti* and *Aedes albopictus* is relatively weak across the country with some notable exceptions, such as Phoenix, AZ, Miami, FL, and Key West, FL. A recent review of funding for mosquito (or insect) abatement indicates vast disparities from Tallahassee, FL—a "high risk" area that spends approximately $23.47 per person—to Jacksonville, FL, which spends only $0.06 per capita [17]. However, even in areas where there are high amounts of funding for mosquito control, these numbers do not differentiate between control for mosquitoes that are merely a nuisance and those that transmit disease so it is unclear the actual amount being spent towards disease control and may not represent what is available in a specific geographic location due to the overlap between county and municipal budgets.

Most surveillance programs were formed to survey West Nile Virus following its emergence in 2003. The techniques used to survey for the *Culex* species that transmit West Nile virus do not directly correlate to *Aedes* species of mosquitoes. They inhabit different ecological niches, and the traps used for *Culex* do not attract *Aedes* species of mosquitoes efficiently. While scaling up of surveillance may not be possible for all jurisdictions, it is critically important for the areas at greatest risk that currently lack good surveillance--chiefly, the U.S.-Mexico border region. Many jurisdictions in the U.S.-Mexico border region, including Yuma and Nogales, Arizona, where I am from, have personnel that are charged not just with vector-control but also restaurant inspections, pest abatement, and other environmental health hazards. This broad range of commitments limits the time that can be dedicated to conducting surveillance and control for mosquito-borne illnesses.

Forecasting systems would need to be targeted to specific geographic areas to provide greater accuracy. Having better surveillance data on *Ae.*

aegypti populations would significantly improve the forecasts of Zika outbreak risks. Integration of this data from multiple sources to validate and improve predictions in real-time is also beneficial. This can come from both traditional surveillance sources and alternative surveillance systems. I am currently involved in an effort to develop a new community-based surveillance "app", Kidenga, along with advisors from the Centers for Disease Control and Prevention (CDC) and Skoll Global Threats Fund. This app will allow smart-phone users to report symptoms that may be consistent with dengue, chikungunya, and Zika viruses and will also allow them to report mosquito activity in their area. The data is aggregated and presented back to the users. Other community-based surveillance activities such as the Great Arizona Mosquito Hunt, led by Dr. Kathleen Walker and partners at the Arizona Department of Health Services, recruit community members to set out simple oviposition traps to collect eggs. The U.S. Department of Agriculture is also encouraging similar participation by community members. Data from these sources can be used to enhance predictions.

We need a better understanding of the relationships between humans and Aedes species to quantify transmission potential. Further, while there is a fairly good understanding of the biological processes of *Ae. aegypti* and *Ae. albopictus* that can be included in forecasting efforts, the human-vector interaction components are still relatively under-studied, particularly in the United States. More information is needed on how frequently *Aedes* species feed on humans in different environments. In addition, understanding how to predict the density of containers and available habitat for *Aedes* species across large geographic regions requires further quantification of the relationship between human demographic factors and *Aedes aegypti* and *Aedes albopictus* indicators.

We need sustained support for model-based forecasts of risk for Zika and related viruses. Partners at NASA and NCAR, including Dr. Cory Morin, Dr. Dale Quattrochi, Mr. Bradley Zavodsky, Dr. J. Brent Roberts and Dr. Andrew Monaghan, are currently working towards this goal but more support is needed to generate these forecasts. We believe that, while imperfect, models have reached a level of sophistication that would enable the provision of actionable information to public health and vector control decision makers about when cities will be at highest risk for virus transmission. Ensembles of seasonal climate forecasts could be used to drive mosquito and virus transmission models to provide forecasts of potential mosquito abundance and virus transmission risk with up to 3 months of lead time for cities across the United States. Issues to address, if such a forecast capability were to be implemented, would include ensuring sustained support (a necessity for any forecasting capability), determining the entity or entities that would operate it, and engaging the public health and vector control communities to maximize the forecast system's utility and iteratively improve it.

We need more information on the resources available for response and control of Aedes species across jurisdictions. To better predict actual risk of transmission, we must also have a better understanding of our capacity to respond. As indicated previously, there is sparse evidence available to develop an understanding of each community's resilience in the face of an outbreak of Zika virus or other mosquito-borne viruses. It is almost certain that many communities, particularly many of those at greatest risk, have little capacity to respond. This information should be obtained in a standardized manner to allow incorporation into the modeling of transmission risk.

CONCLUSION

Our knowledge of *Ae. aegypti* (and *Ae. albopictus*) and our ability to assess present conditions or forecast upcoming risk of Zika transmission and related viruses in the continental United States is currently limited by our incomplete understanding of vector-virus-human interactions and our lack of ground-based surveillance of the geographic and seasonal distribution of both *Ae. aegypti* and *Ae. albopictus*. Sustained support is needed to rectify these gaps including enhancing surveillance and reporting of *Ae. aegypti* and *Ae. albopictus* mosquito populations. Investment in developing and testing forecasting systems is needed. Surveillance and forecasting activities, in particular, need long term stable funding mechanisms to ensure scientific progress. The transition of these activities to operational use is a particular challenge. The timely collection and dissemination of epidemiological and entomological information will be critical for both accelerating research and enabling effective operational programs to forecast and prevent pathogen transmission. These types of investments would improve our capacity to respond, not only to the Zika virus pandemic, but to future threats of viruses that can be transmitted by mosquitoes.

Thank you for the opportunity to testify before you today.

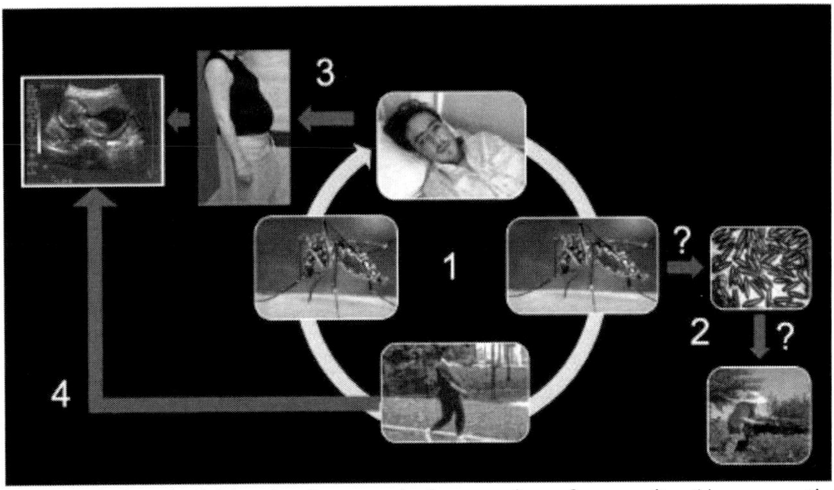

(1) Transmission is initiated when an Aedes species of mosquito (Ae. aegypti is depicted here), feeds on an infected individual. The virus must undergo biological processes in the mosquito before becoming infectious to the next individual. Then that person must go through a period of time before they are infectious to the next mosquito. (2) In addition, it is unknown if infected Aedes can pass the virus on to their offspring. If they can then it would be possible for those offspring to transmit the virus to a human when they first feed. Other modes of transmission not discussed here include (3) sexual transmission, and (4) transmission from a pregnant mother to her fetus. Also possible is transmission through the blood supply and transfusions (not depicted here).

Figure 2. Transmission cycle of *Ae. aegypti* and Zika virus.

REFERENCES

[1] Anderson, R.M. and R.M. May, *Infectious diseases of humans : dynamics and control*. Oxford science publications. 1991, Oxford; New York: Oxford University Press. viii, 757 p.

[2] Weaver, S.C., et al., *Zika virus: History, emergence, biology, and prospects for control*. Antiviral Res, 2016. 130: p. 69-80.

[3] Eisen, L. and C.G. Moore, Aedes (Stegomyia) aegypti in the continental United States: a vector at the cool margin of its geographic range. *Journal of medical entomology*, 2013. 50(3): p. 467-78.

[4] Gubler, D.J., Dengue, Urbanization and Globalization: The Unholy Trinity of the 21(st) Century. *Trop. Med. Health*, 2011. 39(4 Suppl): p. 3-11.

[5] Christophers, S.R., *Aëdes aegypti (L.) the yellow fever mosquito; its life history, bionomics, and structure*. 1960, Cambridge Eng.: University Press. xii, 738 p.

[6] Diallo, D., et al., *Zika virus emergence in mosquitoes in southeastern Senegal, 2011*. PLoS One, 2014. 9(10): p. e109442.

[7] Musso, D. and D.J. Gubler, Zika Virus. *Clin. Microbiol. Rev.,* 2016. 29(3): p. 487-524.

[8] Grunnill, M. and M. Boots, *How* Important is Vertical Transmission of Dengue Viruses by Mosquitoes (Diptera: Culicidae)? *J. Med. Entomol.*, 2016. 53(1): p. 1-19.

[9] Chouin-Carneiro, T., et al., Differential Susceptibilities of Aedes aegypti and Aedes albopictus from the Americas to Zika Virus. *PLoS Negl. Trop. Dis.*, 2016. 10(3): p. e0004543.

[10] Petersen, L.R., et al., Zika Virus. *N. Engl. J. Med.*, 2016. 374(16): p. 1552-63.

[11] *Zika - Epidemiological Update; 21 April 2016*, in *Zika - Epidemiological Update*, P.A.H. Organization; Editor. 2016, PAHO/ WHO: Washington D.C. p. 11.

[12] Faraji, A., et al., *Comparative host feeding patterns of the Asian tiger mosquito, Aedes albopictus, in urban and suburban Northeastern USA and implications for disease transmission. PLoS Negl. Trop. Dis.*, 2014. 8(8): p. e3037.

[13] Monaghan, A.J., et al., On the Seasonal Occurrence and Abundance of the Zika Virus Vector Mosquito Aedes Aegypti in the Contiguous United States. *PLoS. Curr.*, 2016. 8.

[14] Morin, C.W., A.C. Comrie, and K. Ernst, Climate and dengue transmission: evidence and implications. *Environ. Health Perspect.*, 2013. 121(11-12): p. 1264-72.

[15] Magori, K., et al., Skeeter Buster: a stochastic, spatially explicit modeling tool for studying Aedes aegypti population replacement and population suppression strategies. *PLoS neglected tropical diseases,* 2009. 3(9): p. e508.

[16] Morin, C.W., et al., Meteorologically Driven Simulations of Dengue Epidemics in San Juan, PR. *PLoS Negl. Trop. Dis.,* 2015. 9(8): p. e0004002.

[17] Garrett, L., *Springtime in D.C. Means Mosquitoes — and Zika*, in *Foreign Policy*. 2016, Graham Holdings Company.

In: The Zika Virus
Editor: Veronica Salazar
ISBN: 978-1-53610-124-9

Chapter 11

TESTIMONY OF DR. DANIEL NEAFSEY, ASSOCIATE DIRECTOR, GENOMIC CENTER FOR INFECTIOUS DISEASE, BROAD INSTITUTE OF MIT AND HARVARD. HEARING ON "SCIENCE OF ZIKA: THE DNA OF AN EPIDEMIC"*

SUMMARY

The Zika epidemic has caught the world off guard. We lack fundamental understanding of how the virus moves from person to person via mosquitoes, information that is crucial for an effective response to the epidemic. Though the virus was little studied prior to the recent epidemic that began in Brazil, the *Aedes* mosquito that spreads it is better known, because *Aedes* mosquitoes also spread the viruses that cause Yellow Fever, Chikungunya, and Dengue Fever. Despite the mosquito's importance, we lack foundational resources to pursue DNA-based studies of the biology and transmission of Zika. This resource is gap is critical. Infectious disease epidemiology has been transformed by DNA during the last 10 years into a rich digital information science, allowing biologists and public health agencies track the spread of outbreaks over time and space, and learn about what mosquito and human factors contribute to disease spread. We can now tackle emergent infectious diseases like Zika using efficient and innovative genetic tools, and the

* This is an edited, reformatted and augmented version of testimony presented May 25, 2016 before the House Committee on Science, Space, and Technology.

scientific community stands ready to develop and apply these tools to Zika to protect vulnerable populations within our borders and around the world.

RATIONALE FOR DNA-BASED EPIDEMIOLOGY

Insight into the biology of how *Aedes* mosquitoes are able to spread disease can be gained through sequencing and mapping the *Aedes* genome (the entirety of its DNA), by studying the DNA of mosquitoes collected across time and space during the outbreak, and by looking for DNA factors that impact interventions and Zika transmission. DNA sequencing will generate foundational resources that benefit a diverse array of scientific studies, leading to the improvement of existing disease interventions, the preservation of their efficacy, and the discovery of new interventions.

It is exceedingly important that we learn more about the nature of Zika transmission, and identify and invest in new approaches to block disease transmission. We do not yet have a drug to treat Zika patients, nor vaccines with which to protect people from infection.

Our primary disease intervention option at the present time is to stop Aedes mosquitoes from spreading the disease. It is unlikely that we will ever eliminate *Aedes* mosquitoes within our own borders using conventional mosquito control measures, and much of the world lacks the socio-political infrastructure to even mount such an effort. Because the mosquitoes are here to stay, we must learn how to disrupt their biology or strategically suppress their populations in a way that impacts Zika transmission.

It is not a coincidence that Zika is spread by the same *Aedes* mosquitoes that transmit Yellow Fever, Chikungunya, and Dengue Fever, but a consequence of the remarkable predilection *Aedes* mosquitoes have for feeding on humans, and their capacity to utilize human-created environments for reproduction. Out of almost 4000 mosquito species on this planet, very few have assembled the combination of molecular, behavioral, and ecological traits that make *Aedes* such an effective disease vector.

DNA sequencing of mosquito genomes has already given us a glimpse into some of the biological traits that favor human disease transmission by *Aedes*, *Anopheles,* and other mosquitoes [1–3]. We can now appreciate mosquito-born disease transmission as a product of complex biological interactions between a mosquito, the pathogen, and humans.

Because Zika is a new and understudied disease, knowledge about the details of these interactions remains thin for this virus.

There are many ways to translate knowledge of mosquito/human/pathogen interactions into disease control measures. Examples include:

1. Insecticide resistance detection and surveillance

Insecticides can be very effective at controlling mosquito populations. The historic elimination of Yellow Fever, Malaria, and other mosquito-borne disease from the US can be attributed to mosquito control. Global Malaria mortality fell by 60% between 2000 and 2015, in large part due to the intensive use of insecticide-treated bednets and indoor insecticidal spraying [4].

The routine consequence of intensive insecticide campaigns is resistance. Scientists are concerned that further gains against Malaria will be limited by the high prevalence of insecticide resistance in many regions of Africa [5]. Resistance to many insecticides is already locally common in *Aedes* mosquitoes [6, 7]. The genetic factors behind this resistance are not well understood. DNA-based studies comparing resistant mosquitoes versus susceptible mosquitoes can provide indicators of resistance to use in surveillance programs. Similar to strategies being used to limit antimicrobial resistance in bacterial pathogens, we can maintain insecticide efficacy for longer through rational, judicious use of insecticides, informed by DNA-based markers to track the origin and regional spread of resistance.

2. Mosquito population studies

Using DNA to understand which mosquito populations are connected vs. isolated can tell us much about the Zika epidemic. By comparing the DNA of mosquitoes sampled from different locations, we can address a wide range of fundamental questions. For example: Is the geographic spread of Zika due solely to human movement, or do mosquitoes also play a role? Which mosquito populations harbor high levels of insecticide resistance? Will resistance spread from one region to another? Our DNA-based understanding of *Aedes* mosquito populations is very limited compared to *Anopheles* Malaria mosquitoes, where such information has illuminated many aspects of disease transmission and spread. For example, this information has influenced our understanding of how mosquito populations rebound from control programs and restore themselves after dry seasons, and which mosquito populations are smaller (and therefore ripe targets for control) [8]. A recent small-scale DNA study has determined that *Aedes* mosquitoes have established a year-round population of mosquitoes in the Capital Hill neighborhood [9]. Where else are new *Aedes* mosquito populations establishing themselves, and what is their source?

3. Genetic modification of the mosquito

The prospect of using genetically modified mosquitoes to control disease spread is no longer science fiction. For example, Hadyn Parry's testimony will likely describe the Oxitec method of inserting 'self-limiting' genes into mosquitoes that are released in large numbers, to achieve sharp reductions in local mosquito populations.

Recently, a powerful new DNA tool has arrived with the potential to translate biological insights into control measures very quickly. The CRISPR-*Cas9* DNA editing system gives us unprecedented power to insert, turn off, or modify genes in virtually any organism, including mosquitoes. When coupled with a 'gene drive', another DNA-based tool that can rapidly spread a CRISPR mutation through a mosquito population, we have the power to modify wild mosquito populations to restore sensitivity to insecticides, or to bite other animals instead of humans, or to kill pathogens they ingest during a blood meal and not transmit them to the next person they bite. CRISPR gene drives have been demonstrated to be viable in the laboratory with fruit flies [10] and *Anopheles* Malaria mosquitoes [11, 12]. CRISPR gene drives are an example of the increasingly direct connection between DNA and vector-borne disease control.

Improving the Aedes Mosquito Genome Map

One example of a specific DNA-based resource that will assist the Zika response is an improved genome map for *Aedes aegypti*, containing virtually all of the DNA sequence in long pieces for all of the chromosomes. The current genome map for the *Aedes aegypti* mosquito was a significant advancement when it was first released in 20071, but it was far from perfect. DNA sequences are long strings of nucleotides, or 'letters': A, T, G, and C. At 1.5 billion nucleotides long, the *A. aegypti* genome is far larger than that of many other mosquitoes, and much of that extra length is composed of repetitive DNA sequences, (eg ATC-ATC-ATC-ATC ...). Most modern DNA sequence is read in small pieces of 100 nucleotides in length. Assembling a non-repetitive genome map from short nucleotide sequences can be a computationally difficult task, akin to putting together a large jigsaw puzzle without a picture to look at as a guide. Assembling a large, *repetitive* genome from short sequencing reads is more like putting together a jigsaw puzzle with no picture, *where most of the pieces have the same color and similar shape.* As a result, the existing reference genome assembly for *A. aegypti* is in over 36 thousand pieces (despite having only 3 chromosomes), and is estimated to be missing as many as 20% of the genes. Furthermore, there is evidence that some of the assembled regions have been put together incorrectly.

While this genome map has led to some important breakthroughs in understanding the vectorial capacity of *A. aegypti*, we can do better. *DNA sequencing technology has improved in quality and price dramatically since 2007, making it more than 10,000 times cheaper to sequence a nucleotide of DNA now compared to then.* In recognition of this opportunity, Dr. Leslie Vosshall at Rockefeller University organized the *Aedes* Genome Working Group (AGWG) in January 2016, to coordinate efforts in the vector genomics research community aimed at improving the genome assembly map [13].

With no central funding, and organized via Twitter, this upstart group has made rapid progress exploring a wide range of approaches to employ new sequencing technology in pursuit of an improved *Aedes* reference genome. For a price tag of a few hundred thousand dollars, this group aims to produce a much improved version of the genome map released in 2007 at a cost of $18 million. Examples of the new technologies that will enable this feat include:

1) DNA sequencing machines that produce very long strings of nucleotides.

Pacific Biosciences, a member of the AGWG, manufactures sequencers that can read DNA in strings of more than 10,000 nucleotides, far longer than the 700 nucleotide strings produced for the 2007 genome map. To return to the puzzle analogy, this is equivalent to increasing the size of the puzzle pieces and reducing their number.

2) Machines and techniques that tell us which pieces of sequenced DNA belong to the same territory in the map.

Three biotechnology companies (10X Genomics, BioNano Genomics, and Dovetail Genomics) have been recruited to the AGWG because their technology can inform which DNA sequences are proximal to each other on chromosomes. Application of these tools will mean that genome map puzzle pieces are no longer randomly shuffled in one big pile, but organized into small piles of pieces that correspond to different regions of the puzzle.

3) More sophisticated software to assemble the map.

Compared to 2007, there are now many more tools available to put together genome maps. The previously mentioned AGWG corporate partners have their own software to apply their respective data types to the task of genome assembly. This software is often optimized for the attributes of the human genome, however, and may be inexpert at combining different, complementary data types in order to produce the best possible map. Work at academic research institutes in the US to explore the optimal computational use of the new data is therefore a vital component of the AGWG, and is being carried out by investigators with key expertise in several locations, including Rockefeller University, Yale University, the National Human Genome Research Institutes, the University of California San Francisco, Virginia Tech, and the Broad Institute. To use the puzzle metaphor a final time, this advance is equivalent to producing a robot that can scan a field of pieces and quickly find the missing piece to fill a gap.

The AGWG is still in the process of evaluating and optimizing these technologies. The group aims to produce a new genome map for *Aedes aegypti* by late summer of 2016.

Finding and labeling all of the genes in the new map will take additional effort, and relating the function of those genes to insecticide resistance and other mosquito traits influencing disease spread will take years of work from dedicated members of the vector research community. This technologically opportune project to produce an improved DNA map will not only be a foundational investment in the control of Zika, however, but also Yellow Fever, Dengue, Chikungunya, and future emerging viral diseases likely to hitch a ride between hosts via *Aedes* mosquitoes.

CONCLUSION

The work of the AGWG represents both a triumph and a tragedy. It is a triumph because of the open and collaborative spirit of the endeavor, and the rapidity with which the community has responded to address the Zika epidemic. It is a tragedy because it took a public health emergency like the Zika epidemic for the research community to remedy the poor quality of a foundational resource for understanding disease transmission.

Though the AGWG was founded without a solid base of funding, the maturation of its efforts will require focused investment of resources to ensure the quality and integrity of the new genome map it hopes to deliver. Innovative and effective disease control efforts that make use of the improved *Aedes* genome map are an exciting prospect, but will also require committed investment. *The Zika epidemic can become a proving ground for the power of new DNA-based epidemiological and intervention tools. We have the opportunity to demonstrate to the world how new technologies will let us understand, anticipate, and control the spread of an epidemic, and we have an obligation to vulnerable populations to seize this opportunity.*

REFERENCES

[1] Nene, V. et al. Genome sequence of Aedes aegypti, a major arbovirus vector. *Science* 316, 1718–1723 (2007).

[2] Neafsey, D. E. et al. Mosquito genomics. Highly evolvable malaria vectors: the genomes of 16 Anopheles mosquitoes. *Science* 347, 1258522 (2015).

[3] McBride, C. S. et al. Evolution of mosquito preference for humans linked to an odorant receptor. *Nature* 515, 222–227 (2014).

[4] WHO | World Malaria Report 2015. WHO Available at: http://www.who.int/malaria/publications/world-malaria-report-2015/report/en/. (Accessed: 20th May 2016)

[5] Ranson, H. and Lissenden, N. Insecticide Resistance in African Anopheles Mosquitoes: A Worsening Situation that Needs Urgent Action to Maintain Malaria Control. *Trends Parasitol.* 32, 187–196 (2016).

[6] Deming, R. et al. Spatial variation of insecticide resistance in the dengue vector Aedes aegypti presents unique vector control challenges. *Parasit. Vectors* 9, 67 (2016).

[7] Dusfour, I. et al. Multiple insecticide resistance in Aedes aegypti (Diptera: Culicidae) populations compromises the effectiveness of dengue vector control in French Guiana. *Mem. Inst. Oswaldo Cruz* 106, 346–352 (2011).

[8] O'Loughlin, S. M. et al. Genomic signatures of population decline in the malaria mosquito Anopheles gambiae. *Malar. J.* 15, 182 (2016).

[9] Lima, A., Lovin, D. D., Hickner, P. V. and Severson, D. W. Evidence for an Overwintering Population of Aedes aegypti in Capitol Hill Neighborhood, Washington, DC. *Am. J. Trop. Med. Hyg*. 94, 231–235 (2016).

[10] Gantz, V. M. and Bier, E. Genome editing. The mutagenic chain reaction: a method for converting heterozygous to homozygous mutations. *Science* 348, 442–444 (2015).

[11] Gantz, V. M. et al. Highly efficient Cas9-mediated gene drive for population modification of the malaria vector mosquito Anopheles stephensi. *Proc. Natl. Acad. Sci. U. S. A.* 112, E6736–6743 (2015).

[12] Hammond, A. et al. A CRISPR-Cas9 gene drive system targeting female reproduction in the malaria mosquito vector Anopheles gambiae. *Nat. Biotechnol.* 34, 78–83 (2016).

[13] Harmon, A. Team of Rival Scientists Comes Together to Fight Zika. *The New York Times* (2016).

In: The Zika Virus
Editor: Veronica Salazar
ISBN: 978-1-53610-124-9

Chapter 12

STATEMENT OF HADYN PARRY, CHIEF EXECUTIVE OFFICER, OXITEC. HEARING ON "SCIENCE OF ZIKA: THE DNA OF AN EPIDEMIC"*

Chairman Smith, Ranking Member Johnson, and members of the Committee thank you for inviting me to testify before your Committee today. My name is Hadyn Parry and since 2008, I have served as the CEO of Oxitec, a pioneer in using genetic engineering to control inspect pests that spread disease and damage crops. Oxitec was founded as a spinout from Oxford University and is a UK based subsidiary company of Intrexon Corporation, a New York Stock Exchange listed company which engineers biology to help solve some of the world's biggest problems.

BACKGROUND

Oxitec Limited has pioneered the use of bioengineering to provide a solution for controlling the mosquito *Aedes aegypti* that spreads Zika virus. Oxitec developed its product OX513A, a self-limiting strain of *Aedes aegypti* in 2002 and since that date has conducted rigorous indoor and then outdoor evaluation and development. OX513A, therefore, has 14 years of data to support efficacy, environment and safety aspects. In all efficacy trials to date,

* This is an edited, reformatted and augmented version of a statement presented May 25, 2016 before the House Committee on Science, Space, and Technology.

Oxitec has demonstrated a reduction in the target *Aedes aegypti* population by over 90% in about six months. Over recent years, Oxitec has placed considerable focus and investment on its ability to scale up, supply and distribute its insects and OX513A should now be considered a fully operational solution, enabled to be deployed.

OX513A received a preliminary Finding of No Significant Impact (FONSI) from the Food and Drug Administration (FDA) in 2016 following an Investigational New Animal Drug (INAD) filing that was initiated in 2011. The public comment period ended on May 13, 2016 and a final opinion from the FDA is awaited before Oxitec has the regulatory approval to carry out a small trial in the Florida Keys as part of its application. We are hopeful that, in light of the public health need, FDA will act quickly to finalize its finding.

OX513A has national biosafety approval from the National Biosafety Technical Commission (CTNBio) for use in Brazil (April 2014) and Oxitec has been informed that we will shortly receive special registration from ANVISA, an agency within the Ministry of Health, to enable availability in Brazil.

OX513A has a specific recommendation from the World Health Organization (WHO) Vector Control Advisory Group for pilot deployment under operational conditions. WHO rarely recommends a specific product, but took this step as part of its emergency response and preparedness for the Zika virus.

Oxitec's Bioengineering Technology and Status

A) Aedes Aegypti

While the current urgent threat is the Zika virus, the real target is the mosquito that carries the disease. Despite the present and widespread use of insecticides, over the last 50 years, there has been a sharp increase in both incidence and number of diseases spread by *Aedes aegypti* across the world. Before 1970, only nine countries had experienced severe dengue epidemics. The disease is now endemic in more than 100 countries in the WHO regions of Africa, the Americas, the Eastern Mediterranean, South-East Asia and the Western Pacific. WHO reported an estimate of over 390 million dengue infections per year. Chikungunya came into the Caribbean in December 2013 and within one year, there were over one million cases across the Central America and Caribbean region.

By focusing on the mosquito, rather than the disease, Oxitec has accumulated years of development, well ahead of the Zika threat emerging. It also should be noted that it is unlikely that Zika will be the last disease transmitted by *Aedes aegypti*. It is clear that new tools are needed control this vector to guard against not only the current threat but also known threats such as dengue and Yellow Fever along with future, unidentified threats.

Aedes aegypti is the prime vector for Zika virus as well as for dengue, chikungunya and Yellow Fever. Understanding and then controlling the vector is key to controlling the spread of Zika or any other current or future virus transmitted by the mosquito. This mosquito species originated from Africa and has been spread around the world by human activities. This distribution has occurred via the transport of eggs, which are highly durable and easily carried in freight. Unlike many mosquito species which are adapted to the rural environment, this mosquito has adapted to be predominantly an urban one. It bites humans by preference and lives in and around the home and other areas where people congregate. Its larvae and pupae develop in still water pools such as unused containers, birdbaths and blocked gutters. This mosquito can breed in as little as a bottle cap of water.

It is notoriously difficult to control with conventional insecticides not only because it has developed resistance to the most common insecticides but mainly because containers create multiple breeding sites in the urban setting that are often on private property and inaccessible to mosquito control staff. There are just too many dispersed inaccessible breeding sites that need to be continuously treated for this mosquito to be controlled through application of chemical insecticides.

Oxitec has used some key features of this mosquitoes' biology to design our solution. First, only females bite. Males cannot bite and, therefore, males represent no threat to humans.

Second, males are extremely effective at finding females. They tune in to the sound of the female wing beat to locate the female in order to mate. Once the female has mated successfully, she does not need to mate again and she will engage in a cycle of biting humans (to acquire blood), laying eggs, then biting again and will continue on this cycle. A single female can lay up to 500 eggs in her lifetime, which develop from egg to larvae to pupae to adult in a little over one month.

Third, it is important to note that neither male nor female mosquitoes have a significant spontaneous flight range. An adult mosquito will only fly up to about 200 yards in its lifetime.

B) Oxitec's Approach

To control *Aedes aegypti,* Oxitec uses the mosquito against itself. We release males (that cannot bite) that mate with wild females. The offspring inherit self-limiting genes and die before becoming functional adults, thereby reducing the wild population. A male is biologically tuned to seek out females for mating purposes. The wild females cannot distinguish between an Oxitec OX513A male and a wild one, meaning that, provided sufficient Oxitec males are released across an urban area for a period of time, the population will rapidly decrease. In trials in several countries, we have shown that the population of *Aedes aegypti* may be reduced by over 90% in around six months. Moreover, as this species cannot fly far, the effect of the control can be highly specific to the area of release. Specifically, we can treat broad urban areas by releasing in a pattern to cover a town or city, and we can modify the release rate to concentrate more on areas of higher mosquito population. Or we can even specifically target "hotspots". Once controlled, low level releases can be continued in areas of likely re-infestation to sustain the control over the longer term. Unlike chemical insecticides, the OX513A mosquito affects *only* the *Aedes aegypti* species, and has no measurable impact on other mosquito populations, or the overall insect population of the treated region.

C) Oxitec's Technology

There are two key elements to the Oxitec bioengineering approach.

a) *A self-limiting gene.* Each released OX513A insect carries two alleles of the self-limiting gene meaning that each of its offspring inherits one copy. That single copy prevents the offspring from developing to become a functioning adult, so the offspring of a mating between an OX513A mosquito and a wild one die in early development. Oxitec uses the term self-limiting; meaning offspring will die before becoming functional adult. Therefore, these mosquitoes do not reproduce.
b) *A fluorescent biological marker.* In addition to inheriting the self-limiting gene, all of the offspring inherit a fluorescent marker. This gene allows us to identify all the larvae. When viewed under a filter, the larvae show a distinct red color and pattern. This color provides an unparalleled system for monitoring and tracing the Oxitec mosquito.

While the self-limiting gene is the mechanism to reduce the population of Aedes aegypti and, hence reduce the threat of transmission of Zika, the marker allows the program to be precise, efficient and cost-effective. Throughout a program, Oxitec collects eggs from the area of release and determines the proportion of the larvae that have an OX513A parent. More or fewer adult males then can be released in each area. The release rate is tailored to specific requirements and, as a result, overall control of the whole Aedes aegypti population is achieved as swiftly as possible.

D) Programs, Results to Date and Status

An Oxitec program consists of the release of male mosquitoes from predetermined release points in a town or city each week. Global Positioning System (GPS) release points are programmed into a release plan. Male OX513A mosquitoes are released from each GPS point from a moving van. The males then disperse and seek females. Eggs and larvae are collected each week from simple ovitraps (that are commonly used by most mosquito control teams) not only to show the level of Aedes aegypti population but also, using the marker, to adjust the release plan for the following weeks. The overall number of males used, the frequency of release and the number of release points are all determined at the commencement of the program and refined as it continues; the marker allowing for an unprecedented level of control and precision.

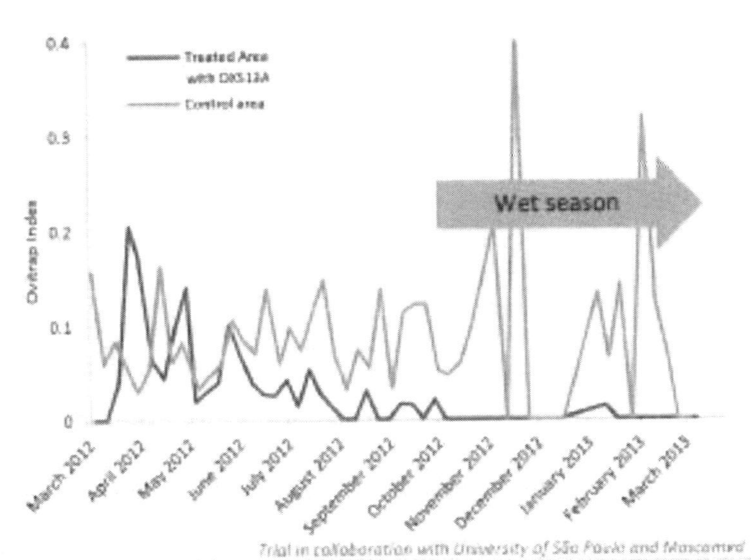

Figure 1. Oxitec's OX513A results in Mandacaru, Brazil suppressed *Aedes aegypti* by >90% and sustained control through rainy season.

Oxitec has demonstrated over 90% reduction in the *Aedes aegypti* population in about six months in all efficacy trials. Figure 1 shows the effect chronologically comparing[1] an area of OX513A release (green line) against a site where no OX513A were released (a control area denoted by the orange line). Following first releases, the population starts to decline rapidly after about three months. As the population declines, releases continue but fewer OX513A males are used as they are not needed at the same level as when the wild population was higher. Even when the wet season starts, the wild *Aedes aegypti* do not recover in the release area with a continuing low level release. By contrast, the untreated area shows a reasonably consistent level of infestation throughout the year with a sharp seasonal rise in the wet season.

Oxitec has placed a major focus on operational preparedness. Oxitec has established permanent production units already in UK and Brazil and we are currently scaling up in Piracicaba, Brazil to a level to supply enough mosquitoes in the next phase to protect up to 1.5 million people. The supply chain can also use features of the mosquito biology to provide efficient logistics and quality control. Eggs can be produced at a central location for an entire country or region, from which they can be stored and distributed as required. More locally, a production hub can receive eggs, rear to pupae, separate males from females and produce the males for release.

E) Safety and Environment

Safety to environment and humans are key aspects of our approach. In stark contrast to insecticides (which will affect a broad array of insect life and respective food chains), Oxitec releases only affect the target mosquito so it is a species-specific approach. *Aedes aegypti* only mate and produce viable offspring with *Aedes aegypti*. Further, this species is an invasive or non-native one in all countries outside of Africa. From a historical perspective, it is a recent invader to the U.S. as well as an urban dweller so native ecosystems have not developed key dependencies. As a food source, it forms a very small part of the diet of other animals. For example, in the Florida Keys *Aedes aegypti* forms less than 1% of the biomass of all mosquitoes.

But perhaps the most important environmental aspect of OX513A is that neither the released adults nor their offspring remain in the environment. Again, it is a self-limiting approach whereby the released males will die after a few days as will their offspring.

Over the last 14 years, Oxitec has conducted a broad array of studies on human safety and the environment and these studies have been used to inform regulatory decision making and these are publically available through the relevant country's regulatory mechanisms. Following the review of this data, the FDA's Center for Veterinary Medicine, working with experts from the Centers for Disease Control and Prevention (CDC) and the Environmental Protection Agency (EPA), has reached a preliminary Finding of No Significant Impact (FONSI). This finding corroborates that of the National Biosafety Technical Commission (CTNBio) in Brazil where OX513A has been approved for releases throughout the country.

Comparison to Other Forms of Vector Control

The main forms of vector control used today throughout the world are a combination of monitoring, good practice (through the prevention or removal of breeding sites though regular inspections and eliminating sources of standing water) and use of insecticides. While insecticides have been the mainstay of vector control products, insecticide resistance and the urban anthropophilic nature of *Aedes aegypti* mean that insecticides are not effective in controlling population of *Aedes aegypti* across an urban environment.

In addition to insecticides, there are a number of new innovations proposed that range from Oxitec's OX513A with proven efficacy along with over a decade of data, to those still in the research phase or suggested for future research. Perhaps the clearest way to differentiate these new innovations is based on

a) whether the released insects are designed to spread and persist in the environment, or not; and
b) field trial evidence and operational preparedness.

Spreading and Persisting Approaches (Population Replacement)

One school of thought relies on the concept of replacing the population of *Aedes aegypti* with a different, modified version. This could be achieved through genetic engineering or through other means. Using genetic engineering, genes may be introduced through the mosquito population (gene drive) with the intent of replacing the existing population in a biased or driven way that is less harmful. There are different early stage gene drive research programs investigating this approach.

Another approach is to try to achieve the same outcome (population replacement) by infecting the mosquito population with a bacterium. Wolbachia is an intracellular bacterium that is not naturally present in *Aedes aegypti* but, once infected with a specific wolbachia strain (wMel), investigators report a reduction in the viral load in the mosquito in laboratory experiments. To date, outdoor trials have concentrated on replacement of the wild *Aedes aegypti* population with the wolbachia (wMel) infected strain. Efficacy trials to examine the disease impact have not yet taken place.

Regardless of how these prototypes are developed, one needs to both replace the existing mosquito population with the modified version but also ensure that the new modified version is significantly less of a threat over the long term than the wild *Aedes aegypti* that it replaces. Both males and females will need to be released in order to allow the modified mosquito to replace the wild version. If the mechanism of action results in a reduction in virus rather than a complete block, then releasing biting females when virus is endemic may actually assist virus spread.

In practice, population replacement is more complex and a potentially riskier strategy than that of use of Oxitec's technology, since population replacement results in a new population of mosquitoes modified either by genetic means or through infection with a bacterium. It is therefore essential to consider the long-term impact of the evolving mosquito combined with the bacteria and combined with virus. A particular focus for products that are designed to spread and persist in the environment is the potential long-term effects as by their nature there is no obvious ability to contain the spread or engage a product recall if this is required. Stewardship will also be a key issue.

Non-Spreading and Non-Persisting (Population Reduction)

Population reduction approaches rest on using products that do not persist or spread in the environment. The product is used for as long as it is required and no longer. Some may argue that long-term control means ongoing releases but in effect we already have ongoing (but inadequate) vector control with chemical products. The ability to suppress populations of disease-carrying mosquitoes without ongoing consequences is a key advantage of these approaches.

Oxitec is a leading proponent of this approach with OX513A. Both the released males and their offspring die, meaning that there is no spreading in the environment and no persistency.

There is also significant precedent in this approach in agriculture through the Sterile Insect Technique (SIT) that was developed using radiation in the 1950's and 1960's. The use of radiation devices can impose limitation on deployment and the damaging nature of the radiation can weaken the mating fitness of the insects. Mosquitoes have been a key target for radiation driven SIT but have enjoyed mixed success over many years. *Aedes aegypti* radiation based programmes have been suggested as a response to Zika but no field trials have yet taken place so the applicability of radiation-based SIT to *Aedes aegypti* is not yet tested or proven.

Field Trial Evidence and Operational Preparedness

A key consideration is that Oxitec commenced its development in 2002. At that time Oxitec's core focus was *Aedes aegypti* due to the unmet need to control this vector with respect to dengue. Even since 2002 the geographic spread of dengue and the number of cases has risen alarmingly. To this has been added the Chikungunya epidemic and now Zika. But this singular focus on the vector has meant that Oxitec's OX513A has accumulated an extensive body of efficacy and other supporting data, especially since first outdoor trials in 2009. Ironically as a genetically engineered organism it has undergone arguably a far higher standard of regulatory and independent scrutiny than non GE approaches.

Also, development has focused on operational practice, namely the development of methodologies, quality control and standards that have arisen as part of the maturation of the program. This level of data and operational use preparation sets OX513A apart from other approaches that are much earlier in research and development cycle.

Regulatory Status with a Focus on the United States

The regulatory environment for all new approaches should be predictable, consistent and rigorous. The innovation that is needed in the area of vector control is stifled when regulatory delays occur. When serious diseases are involved, these delays can have life-altering consequences.

Oxitec's involvement in the United States began in 2009-10 when the Florida Keys experienced local transmission of dengue. The Florida Keys Mosquito Control District (FKMCD) determined that they needed new tools in order to protect their citizens from the mosquito that is the main disease vector: *Aedes aegypti.*

FKMCD sought the opportunity to evaluate Oxitec's technology, and it was agreed to conduct a small scale efficacy trial in a defined area in the Florida Keys to test our technology for the control of *Aedes aegypti.* Oxitec initially applied to USDA-Veterinary Services for permission to conduct the trial. In 2011, USDA-VS determined that it did not have regulatory authority over the Oxitec mosquitoes as it could envisage no risks to animal health from its use. Therefore, in the absence of another regulator, the FDA-CVM assumed regulatory responsibility by regulating OX513A under the new animal drug provisions of the Federal Food, Drug, and Cosmetic Act (FD&C Act). As set forth in FDA Guidance on Genetically Engineered Animals (referred to as "Guidance 187"), the agency regulates genetically engineered animals under its new animal drug authorities on the basis that the recombinant DNA construct introduced into the genome of the animal is intended to affect the "structure or function" of the animal and, thereby, meets the "drug" definition under the Act. As a result, genetically engineered animals are subject to mandatory pre-market approval by FDA.

In late 2011, Oxitec opened an investigational new animal drug (INAD) file with the FDA-CVM for a small scale efficacy trial in the Florida Keys, building on the promising efficacy results from Brazil and Cayman. While the recombinant DNA introduced into Oxitec mosquitoes is regulated by FDA as a new animal drug, bacteria, such as wolbachia, that are introduced into mosquitoes as a form of vector control are regulated by the Environmental Protection Agency (EPA) as pesticides. The impact of this distinction in regulatory jurisdiction is substantial as there are significant procedural differences between FDA's and EPA's approval pathways. The FDA regulatory process for new animal drug approval is rigorous and multi-faceted. Companies seeking FDA approval generally must provide considerable and comprehensive data and information to establish safety and effectiveness, perform an in-depth assessment of environmental impacts, develop drug labeling for FDA approval, and otherwise comply with several pre- and post-market regulatory requirements. In order to assess environmental impacts, FDA-CVM brought together regulators from EPA and CDC as well as other experts for this evaluation - this took a considerable time to form the review team (6-9 months) as the agencies negotiated a memorandum of understanding to govern the interagency consultation. Meanwhile, Oxitec, as the regulated company, was prevented from moving forward with the efficacy trial in the Florida Keys.

FDA-CVM operates a modular dossier submission structure and Oxitec provided the first module in Dec 2013 and the last in Feb 2016, which was the final version of the sponsor-authored draft environmental assessment (EA) prepared in accordance with NEPA requirements and originally submitted to FDA in 2014. No items are outstanding from the FDA-CVM review for the efficacy trial and Oxitec has received letters of adequate response from FDA-CVM on modules submitted. Following this lengthy and time-consuming pre-trial process, FDA published on March 14th 2016 the draft Environmental Assessment (EA) for public comment along with a preliminary finding of no significant impact (FONSI) on human or environmental health. The public comment period was initially set for 30 days and the period was further extended for an additional 30 days at the request of non-governmental groups such as the Center for Food Safety. The public comment period finally closed on May 13th 2016, following the submission of over 2500 public comments. FDA-CVM is now required to review these comments for substantiveness and will prepare a final EA. Thereafter, Oxitec should be able to move forward with the efficacy trial in Florida. Given the urgency of moving forward, we hope that FDA will make the review of these comments one of its highest priorities and finalize their assessment quickly.

In the Keys, both the FKMCD and Oxitec have proceeded with a policy of transparency. The positive for this is that it allows for public information to accrue over a period of time. Interestingly, since 2013, there have been a number of public surveys that -- despite the absence of recurring disease transmission in the Keys -- show a consistent level of strong support for the approach. A recent study by Purdue University reports 78% support respondents for the use of GE mosquitoes as part of the battle against Zika.

There will always be a divergence of opinion for any new intervention and there are those who are against the trial. The Florida Keys Mosquito Control Board has now agreed to hold a referendum in Monroe County to ascertain local opinion on the trial being carried out. This is likely to occur in August 2016.

In Brazil the first trial with the Oxitec mosquito, OX513A, was conducted in 2010 in Juazerio, Bahia, NE Brazil. This was followed by additional trials. These trials demonstrated no adverse effects resulting from the release of the Oxitec mosquitoes. Moreover, the trials generated sufficient data to submit a commercial application to the Brazilian National Biosafety Technical

Commission (CTNBio) in July 2013. In April 2014, CTNBio approved Oxitec's application for use of the Oxitec mosquitoes throughout Brazil.

Oxitec has been informed that we will shortly receive special registration from ANVISA, the Health Agency within the Ministry of Health, to enable widespread availability in Brazil. Oxitec should therefore be able to bring OX513A into use throughout the country to support the fight against Zika.

In Brazil, both dengue and Chikungunya are endemic and Zika has come into the public consciousness with alarm from 2015. Public support for the Oxitec approach has been at a high level- up to 96% support in Piracicaba; the first operational project. The Brazilian press has been almost uniformly supportive and indeed have given OX513A their own name describing it as *Aedes aegypti do Bem* or 'the friendly mosquito'.

Whilst at the time of writing there is no known local transmission of Zika on the United States mainland, local transmission is expected by many experts. However, there is already significant transmission in Puerto Rico and the first microcephaly case.

Figure 2. Brazil press describe Oxitec solution as 'The Friendly Mosquito'.

Considering it has already taken over 3 years due to a lack of clarity of the regulatory process for the analysis of a small scale efficacy trial we urge FDA to do the following:

a) expedite the final review of this dossier;
b) consider using its enforcement discretion to allow for rapid review of future environmental assessments, given the low risk profile of this product; and
c) given the current Zika crisis, grant emergency use or other expeditious regulatory solutions for the widespread approval of this promising vector control tool.

In conclusion, this technology has a very real potential of assisting in preventing and mitigating the Zika public health crisis. Here in the US and globally, I hope that we can work with the Congress and the Administration to expedite the approval and adoption of this promising technology. Our hope is that communities will have meaningful access to this technology in a timely manner as part of an integrated vector control approach.

Mr. Chairman, Ranking Member Johnson, and members of the Committee, thank you for the opportunity to testify today and I look forward to answering your questions and working closely with you in the weeks ahead.

Related Nova Publication

Ebola: Essentials, Response Efforts, and Prevention Issues

MaryAnn Cole

ISBN: 978-1-63482-253-4 Publication Date: 2015

The Ebola virus is named after the Ebola River, near where the virus was discovered in 1976 in Zaire, now known as the Democratic Republic of the Congo (DRC). It is in the filovirus family, so called because of its filamentous shape. EVD is also known as Ebola hemorrhagic fever. The disease sometimes causes hemorrhage (i.e., bleeding) from body openings, but this symptom is not consistent. Five strains of Ebola virus have been identified. This book discusses the essentials, response efforts, and prevention issues involved with the Ebola virus

INDEX

A

B

C

D

G

H

I

J

K

L

M

N

O

P

Q

R

S

T

U

V

W